LabVIEW - Virtual Instrumentation in Education and Industry

Edited by Petru Adrian Cotfas, Daniel Tudor Cotfas and Horia Hedesiu

Published in London, United Kingdom

LabVIEW - Virtual Instrumentation in Education and Industry
http://dx.doi.org/10.5772/intechopen.102279
Edited by Petru Adrian Cotfas, Daniel Tudor Cotfas and Horia Hedesiu

Contributors
Alexandru Băloi, Branislav Hanko, Daniel Tudor Cotfas, Dušan Koniar, Felicia Băloi, Florin Molnar-Matei, Horia Hedesiu, Jozef Šedo, Libor Hargaš, Marek Paškala, Matúš Danko, Ondrej Hock, Peter Šindler, Petru Adrian Cotfas

First published in London, United Kingdom, 2024 by IntechOpen
IntechOpen is the global imprint of INTECHOPEN LIMITED, registered in England and Wales, registration number: 11086078, 167-169 Great Portland Street, London, W1W 5PF, United Kingdom

British Library Cataloguing-in-Publication Data
A catalogue record for this book is available from the British Library

Additional hard and PDF copies can be obtained from orders@intechopen.com

LabVIEW - Virtual Instrumentation in Education and Industry
Edited by Petru Adrian Cotfas, Daniel Tudor Cotfas and Horia Hedesiu
p. cm.
Print ISBN 978-1-83969-063-1
Online ISBN 978-1-83969-064-8
eBook (PDF) ISBN 978-1-83969-068-6

Meet the editors

Petru Adrian Cotfas is a full professor in the Electronics and Computers Department, Transilvania University of Brasov, Romania. He received a BSc in Mathematics and Physics and a BSc in Computer Science in 1997 and 2001, respectively. He obtained an MSc in Mathematics and Computer Science and a Ph.D. in Material Science Engineering from Transilvania University of Brasov, in 1998 and 2007, respectively. Having vast experience in virtual instrumentation, data acquisition, graphical programming, remote engineering, photovoltaics, and hybrid systems characterization and testing, Dr. Cotfas has published 11 books or book chapters and more than 150 papers in international and national journals and conference proceedings.

Daniel Tudor Cotfas is a full professor in the Electronics and Computers Department, Transilvania University of Brasov, Romania. He received a BS and MS in Mathematics and Physics from the same university, in 1995 and 2001, respectively. He obtained a Ph.D. in 2008 and a Habilitation in Electronic Engineering, Telecommunications, and Information Technology in 2019. He is the author of 11 books and book chapters and more than 100 articles. He is an associate editor of the *International Journal of Photoenergy* and *Frontiers in Energy Research* and a guest editor for several other journals. He also holds one patent. His current research interests include renewable energy, energy harvesting, hybrid systems for photovoltaics, optoelectronics, virtual instrumentation, and remote engineering.

Horia Hedesiu received his BSc and Ph.D. in Electrical Engineering from the Technical University of Cluj-Napoca, Romania (formerly Polytechnic Institute of Cluj) in 1991 and 1999, respectively. He currently works as a professor at TUCN in the Electrical Machines and Drives Department. He has written more than 100 publications, including papers, books, conferences, and magazine articles. He holds two US patents in embedded data collecting systems. His research interests are real-time simulation systems, hardware-in-the-loop, and information systems. He also devotes resources to the development of industrial applications involving machinery, graphical programming, and computer-based measurement systems.

Contents

Preface

These days, a vast range of fields, including economics, chemistry, pharmacology, and all forms of engineering, apply the concept of virtual instrumentation (VI). Anywhere that something needs to be measured, monitored, tested, or controlled, VI can be employed. The primary benefit of VI is that its built-in instruments are based on customizable software and modular hardware. The produced instrumentation's software component grows to be quite significant. Without requiring any changes to the hardware, the software can be used to define the functionality, updates, or enhancements of the built instruments. The measuring strategies employed, the algorithms put into practice, the data processing methods, and the graphical presentation of the results can all make all these possible.

Another cutting-edge idea is Graphical System Design (GSD), which enables the quick development of novel systems based on a unified software platform and commercially available off-the-shelf hardware. GSD starts from an idea and moves through the stages of designing, prototyping, and deploying to achieve and put to market a reliable and modular system as soon as possible. Using the right programming environment is essential to speed up the creation of instrumentation. This book discusses the LabVIEW development environment from National Instruments (NI) as a solution for VI and GSD implementation. NI LabVIEW is a graphical programming language where commands are executed by connecting graphical blocks with wires in accordance with the data flow paradigm.

Chapter 1, "Introductory Chapter: An Overview of Using Virtual Instrumentation", presents the ideas behind VI, GSD, and LabVIEW's ability to put these ideas into real applications. This chapter also presents an overview of LabVIEW, including what it is, how it is used, and where it is utilized, based on bibliometric analysis.

Chapter 2, "Using of Virtual Instrumentation in the Teaching of Autotronics", presents a method for using LabVIEW to create instructional aids for teaching autotronics The chapter explains how the NI cRIO platform and LabVIEW can be used to comprehend the CAN bus and CAN protocol. The first section of the chapter presents the purpose behind this development, followed by key concepts regarding the operation of the CAN bus and an example of how LabVIEW and NI cRIO can be used to analyze and process CAN messages. The final section of the chapter provides a demonstrator for real-time control and communication on a CAN bus.

Chapter 3, "Virtual Instrument for Single-Phase Control of Three-Phase Reactive Power Capacitive Compensators", highlights an electrical engineering solution that may be used in both educational and industrial settings. To increase the power factor in the transmission and distribution facilities, the chapter suggests a virtual instrument created with NI LabVIEW and the NI cRIO platform. This instrument is designed to monitor a three-phase load and manage a three-phase compensator. The opening section of the chapter presents a brief theoretical discussion of power factors and reactive power's capacitive compensation. The second section of the chapter discusses the utilized hardware, an NI cRIO platform with specialized I/O modules, the created LabVIEW application for the suggested automatic compensation algorithm, and the outcomes.

Chapter 4, "Automated Measurement of Traction Line Parameters for Railways", presents a method for measuring and monitoring key traction line parameters to improve the safety of railway transport systems. This chapter examines the characteristics of how the traction line height and stagger can be observed and calculated in practical situations. The measuring system comprises a self-developed electronic device connected to the necessary sensors and a LabVIEW application that enables data processing, graphical user interface presentation, and communication between the PC and the electronic device. The second section of the chapter presents the measurement methodology. This is followed by a description of the acquisition system, which uses a 32-bit NXP microcontroller. Finally, a PC version of the LabVIEW application is presented, allowing users to configure the electronic device, take measurements online, and analyze (post-process) the collected data.

Chapter 5, "Virtual Instrumentation Used in Renewable Energy", discusses the use of NI LabVIEW for the development of instrumentation in the renewable energy domain, ranging from very basic applications to more complex ones created for field research, before concluding with an industrial application. The chapter begins by introducing some theoretical concepts related to photovoltaic panels and thermoelectric generators. It then demonstrates simulation solutions based on these concepts using LabVIEW with the LabVIEW Control Design and Simulation toolkit, NI LabVIEW Multisim API toolkit, and NI Multisim SPICE Simulator. The second section describes applications for measurements starting with those based on basic data acquisition boards and moving on to those based on the NI ELVIS and NI cRIO platforms. The final section of the chapter includes a real-world application for monitoring a home solar system that was created using the NI cRIO platform and the Industrial Internet of Things (IIoT) concept.

The book is dedicated to all those who want to understand the VI and GSD concepts implemented using NI LabVIEW through some real study cases from different domains like education in automotive or electrical engineering, rail transportation, and renewable energy.

The editors would like to express their gratitude to the chapter authors for their outstanding work in crafting the best version of the text that addresses their particular area of expertise, as well as for their cooperation in patiently addressing and incorporating editorial recommendations and feedback. We would also like to thank IntechOpen for providing us with the opportunity to publish this book and share our ideas and work.

Petru Adrian Cotfas and Daniel Tudor Cotfas
Faculty of Electrical Engineering and Computer Science,
Electronics and Computers Department,
Transilvania University of Brasov,
Brașov, Romania

Horia Hedesiu
Faculty of Electrical Engineering,
Electrical Machines and Drives Department,
Technical University of Cluj-Napoca,
Cluj-Napoca, Romania

Chapter 1

Introductory Chapter: An Overview of Using Virtual Instrumentation

Petru Adrian Cotfas, Daniel Tudor Cotfas and Horia Hedesiu

1. Introduction

1.1 Virtual instrument concept

We live in times when technology evolves very quickly in all domains, from medicine to automation or computer science. From measuring and controlling local systems and storing data in local PCs or company servers to measuring and controlling widespread distributed systems, cloud computing, and cloud storing data, the used technologies become more complex and require more effort to understand and use them.

Even though the concept of virtual instrumentation is not a new one, being introduced and used in the late seventies, it can be used with great success to facilitate the use and development of new technologies. One of the most widely used definitions states that virtual instrumentation is a combination of modular hardware and customizable software dedicated to making complex user-defined measurement, test, and control instruments. The instrument obtained based on the virtual instrumentation is called a virtual instrument (VI).

The evolution of computers and embedded systems also affects the evolution of virtual instrumentation due to the fact that the functionality of a device can be defined and modified by software and not totally by hardware. Based on this concept, the goal of a device can be modified by changing/updating only the software components (e.g., an oscilloscope that includes a data acquisition board and a software application can be changed in a spectral analyzer by adding a suitable data processing algorithm, or can be remotely controlled based on a network protocol and a suitable software application).

National Instruments, now NI, introduced the VI concept in the early eighties, having three components: Acquisition> > Processing > > Presentation. If the first component is more based on hardware, the last two are more based on software, which include data processing techniques (implemented in software) and data presentation facilities based on graphical user interfaces. By adding new data processing techniques, GUI components, or communication facilities, more complex and powerful instruments are obtained. On the other hand, sometimes not all the facilities of a traditional instrument (TI) are useful, being difficult to be used, and so, simpler instruments are recommended. Therefore, based on software, only the required facilities can be implemented to obtain a simple, user-friendly, and reliable instrument. Dedicated processors for specific tasks were an advantage for TI not so long ago, but now, the use of multicore processors and RT operating systems for VIs allows

IntechOpen

this advantage to be mitigated or even overcome [1]. Meanwhile, other advantages for using VI appeared. The virtual instrument performance can be increased by upgrading the PC, or the used embedded system or parts of it (processors, RAM memory or storage memory, communication interfaces, etc.), or by upgrading the version of the used software. Thus, by replacing parts of the VI, not entirely, its performances can be increased. This advantage can be a part of the VI benefits versus TI related to the: price versus performance, maintenance, customization, and flexibility of usage depending on the user needs.

1.2 Graphical system design concept

A more recent concept has been used in instrumentation development, namely, Graphical System Design (GSD). According to Wikipedia [2], the GSD is "a modern approach to designing measurement and control systems that integrates system design software with commercially available off-the-shelf (COTS) hardware to dramatically simplify development." According to NI [3], the GSD is "a revolutionary approach to solving design challenges that blends intuitive graphical programming and flexible commercially available off-the-shelf (COTS) hardware to help engineers and scientists more efficiently design, prototype, and deploy embedded systems." Through this GSD concept, NI suggests using a unified software platform for the three stages of product implementation, namely, designing, prototyping, and deployment.

The unified software platform should allow the implementation of algorithms and mathematical models developed theoretically for the simulation of the desired system under the actions of stimuli and constraints that are obtained through theoretical approach or measured on real systems and stored in databases. This represents the design stage.

The second stage supposes building the system prototype at the laboratory level based on the same software platform and COTS hardware for project validation. The prototype is tested under action of the stimuli and constraints that are obtained at the laboratory level.

The third stage involves implementing the application on the final hardware platform used through appropriate adaptations and sizing. The final obtained system is tested under the stimuli and constraints that are obtained from the real working environment.

The time to implement a system based on a unified software platform is shortened because all parties involved have a good knowledge of it, and the platform also provides a consistent API that allows the control of modular and specific hardware.

2. NI LabVIEW as VI and GSD implementation environment

There are many solutions on the market that allow the implementation of the abovementioned concepts, VI and GSD. Such a unified software platform is a graphical programming platform, named NI LabVIEW (Laboratory Virtual Instrumentation Engineering Workbench). This platform offers modern programming techniques, such as graphical programming based on data flow paradigm, project-based programming, object-oriented programming, parallel and distributed programming optimized on modern multicore processors, and so forth. Being focused on the engineering field, there are many facilities, add-ons, and libraries that allow developing application for instruments control, monitor and testing systems, modeling and simulating, economical analyzing, digital signal and image processing, remote control, and web services. The last two mentioned facilities open widely the possibilities of developing applications in the field of Internet of Things, Industrial Internet of Things, and Cloud Computing.

NI LabVIEW is a graphical programming language that is different from other well-known programming languages. The big difference between these programming languages is that the programming in LabVIEW supposes the interconnection of graphical blocks with wires. The file extension of the LabVIEW applications is *.vi*, which comes from virtual instrument. The LabVIEW application consists of two windows: the first, called Front Panel (**Figure 1a**), represents the user interfaces, and the second, called Block Diagram (**Figure 1b**), represents the source code.

LabVIEW offers a large number of functions that allow implementing any kind of application in the fields of measuring, monitoring, testing and control, and more. LabVIEW includes advanced functions for data acquisition, signal processing, instrument communication, image processing, mathematics, data saving, control design and simulation, synchronization, report generation, and so on.

The advantages of LabVIEW programming for testing and monitoring applications include:

- easy and fast development of professional user interfaces (UI) through drag-and-drop UI objects that allow setting and displaying control values and results, respectively;

- connecting thousands of devices and instruments developed by NI or third parties using built-in functions;

- reusing already developed IP written in other programming languages such as Python, C/C++, MATLAB,. NET, VHDL, and more;

- "Programming like you think" based on the data flow paradigm.

In order to increase the flexibility, LabVIEW offers the possibility of interoperability with other applications or programming languages, such as SQL database based on *Database Connectivity Toolkit*,. NET environment based on. *NET functions,* MATLAB

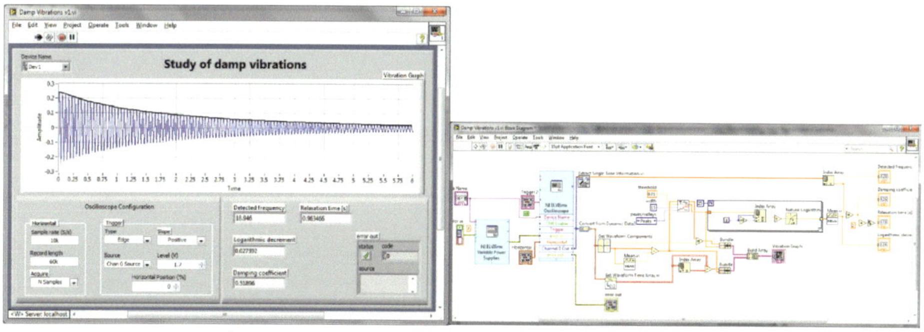

Figure 1.
The front panel (a) and block diagram (b) windows of a LabVIEW application.

based on the *MATLAB script nodes*, Python code based on *Python functions*, SPICE simulation based on *Multisim API*, DLLs through *Call Library Function Node*, and so forth (**Figure 2a**). On the other hand, LabVIEW allows implementing and using protocols for network or bus communications, such as Modbus, TCP/IP, UDP, FTP, HTTP, Serial, Bluetooth, DataSocket, Network Shared Variables, Network Streaming, and so on, or cloud communications based on SystemLink API (**Figure 2b**).

The concept of "programming like you think" correlates well with the data flow diagram used to graphically represent the algorithms that should be implemented in a programming language. Using graphical programming in LabVIEW, the data flow diagram is naturally transposed into code. This transposition is shown in **Figure 3**, where one can see how the initialization, measurement, processing, action, closure, and data saving steps are implemented. At the same time, parallel threads are easy to implement, through parts of the code that are not connected by threads.

Different programming architectures could be implemented in NI LabVIEW to increase the complexity of the developed applications. The developed application should respect the SMoRES concept, which supposes designing a code that has a high Scalability, Modularity, Reusability, Extensibility, and a solution implementation as Simple as possible. Such programming architectures are:

- State machine architecture—allows designing different states of the application that can be accessed any time based on the results of code execution.

- Producer-consumer architecture—dedicated to asynchronous multithreaded applications and where the consumer thread is executed only when the consumer data is available.

- Master-slave architecture—where the main thread always runs and triggers the slaves when to start executing code. The slave executes the code until it is completed and then waits for another trigger.

- Queued state machine and event-driven producer-consumer architecture—a combination of the other architectures used to implement a more complex application that allows handling multiple user-generated events stored in a queue and

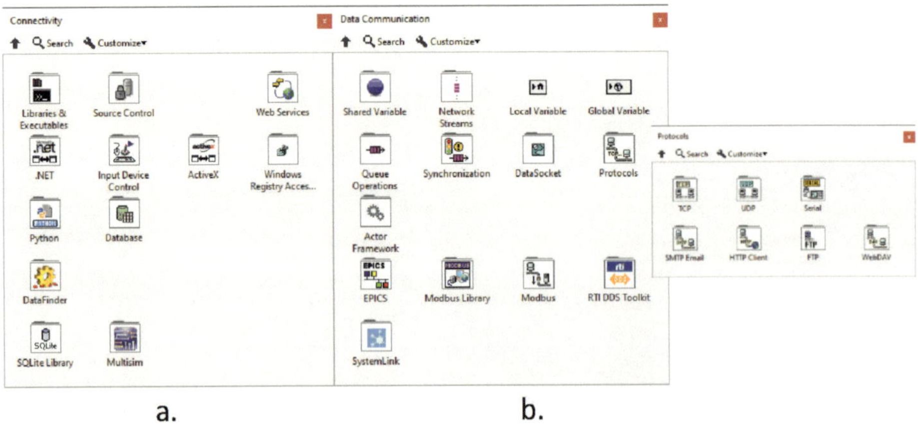

Figure 2.
a. LabVIEW connectivity pallet; b. LabVIEW data communication pallet.

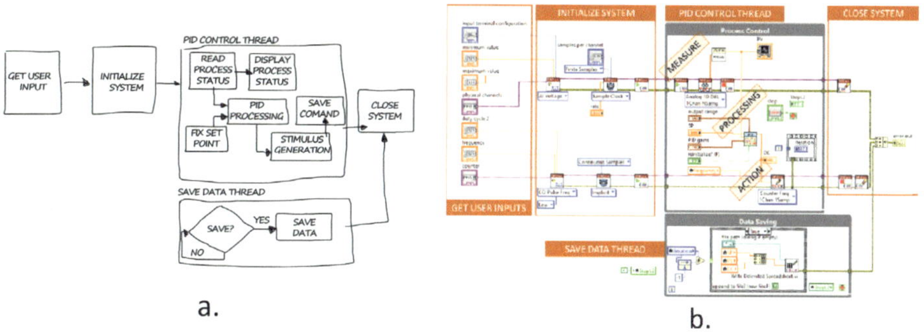

Figure 3.
The data flow diagram (a) transpose in NI LabVIEW code (b).

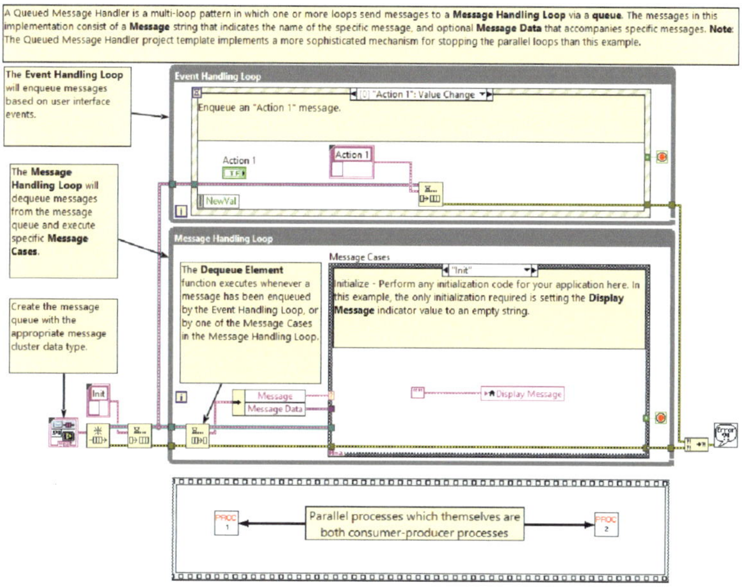

Figure 4.
LabVIEW example of queued state machine & event-driven producer-consumer architecture (from LabVIEW example library).

controls the sequence of states in the state machine consumer thread [4]. Other consumer threads can consume events generated by the state machine consumer, as can be seen in **Figure 4**.

LabVIEW can be used in industry, education, and research domains. A simple search into the scientific databases shows that LabVIEW is widely used in the research domain. Using "LabVIEW" as a search term, the IEEE Xplore database returns 6062 results divided into the following categories: conferences—5663, journals—342, magazines—40, books—13, early access articles—4, and ScienceDirect database returns 45,850 results, which are spread over the following domains: Engineering—17,092, Materials Science—9535, Physics and Astronomy—7925, Chemistry—7847, Energy—7168, Chemical Engineering—6650, Medicine and Dentistry—5422, Biochemistry, Genetics and Molecular Biology—3961, Neuroscience—3119, and Environmental Science—2375.

In the case of Scopus database, the search returned 36,561 results spread on the following domains: Engineering—23,373, Computer Science—13,193, Physics and Astronomy—8208, Mathematics—4789, Materials Science—4314, Energy—3812, Social Sciences—1968, Medicine—1673, Chemistry—1543, Chemical Engineering—1444, Biochemistry, Genetics and Molecular Biology—1335, Environmental Science—1194, Decision Sciences—1164, Agricultural and Biological Sciences—960, Earth and Planetary Sciences—815, Business, Management and Accounting—507, Multidisciplinary—371, Neuroscience—312, Health Professions—241, Pharmacology, Toxicology and Pharmaceutics—139, Arts and Humanities—125, Psychology—114, Immunology and Microbiology—92, Economics, Econometrics and Finance—80, Veterinary—31, Dentistry—30, and Nursing—26. The IntechOpen Publisher contains a list of over 20 chapters or books that include the term LabVIEW, since 2010. These results[1] spread on a very large spectrum of domains show that LabVIEW can be used with success in applications that suppose measurement, testing, simulation, and control, even if these fields are not among the engineering ones. Considering the above results, some domains of using LabVIEW are shortly presented below.

2.1 NI LabVIEW and embedded systems

The last development imposes the usage of open embedded systems, such as Arduino or Raspberry Pi platforms. Therefore, add-ons or APIs were developed for using these platforms with LabVIEW as it is shown in [5, 6]. Moreover, there is an add-on, named Arduino-Compatible Compiler for LabVIEW developed by the TSXperts company, that allows to compile and deploy VIs developed in LabVIEW on the Arduino platforms. For more complex applications based on embedded platforms, the NI LabVIEW RT and LabVIEW FPGA can be used [7, 8]. By implementing metaheuristic algorithms in FPGA, one can increase the power of the developed application [9]. In the case of complex applications based on embedded systems, NI considers one device that includes a processor, an FPGA, and a modular input/output (I/O). The architecture of this device is known as reconfigurable I/O (RIO) architecture. Using the NI LabVIEW FPGA module, the FPGA can be programmed, and using the built-in connectivity functions, the I/Os and processor can be accessed. With the LabVIEW RT module, a deterministic hard real-time application can be developed for control, data processing, and network connection with additional systems. The ease of writing FPGA code in LabVIEW FPGA is given by the fact that it is the same as writing normal LabVIEW code, using only the functions available for FPGA (**Figure 5**).

2.2 NI LabVIEW in energy

The energy demand of humanity is increasing, and therefore, the new energy sources or improving the efficiency of the existing ones are studied. Consequently, the models and simulations, testing and controlling, measuring, and monitoring of these sources should be done at industrial and educational levels. The LabVIEW Electrical Power Toolkit is one tool included in LabVIEW that offers the possibility to develop applications for electrical power and power quality measurements on hardware platforms like CompactRIO, CompactDAQ, and PXI. Characterization tools

---

[1] The results were obtained in 08.10.2023.

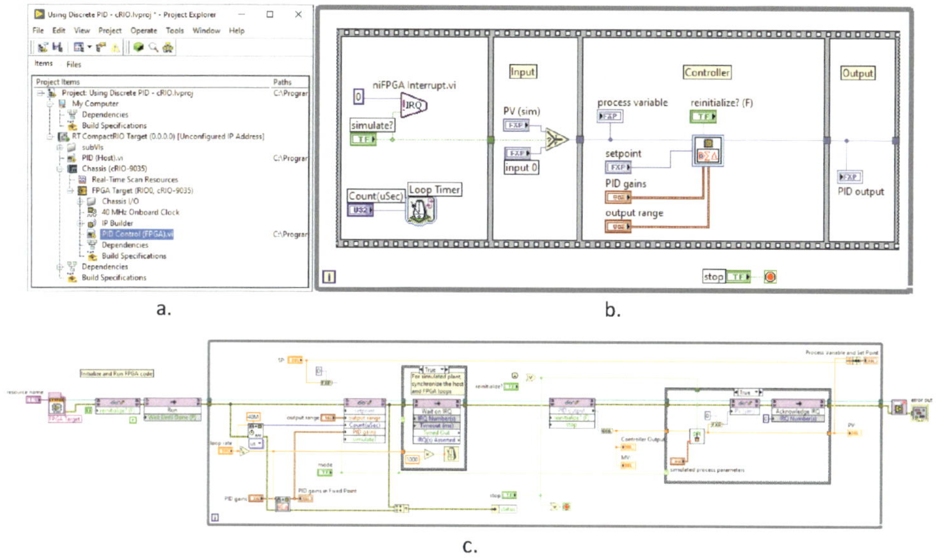

Figure 5.
NI LabVIEW RIO architecture implementation (from LabVIEW example library). a. LabVIEW project with PC, RT, and FPGA branches; b. LabVIEW FPGA implementation; c. LabVIEW RT implementation.

based on VI are often used in the field of photovoltaic systems [10–12], wind turbines [13–15], or hybrid systems [16–18].

2.3 NI LabVIEW in automotive industry

Test and control systems are one of the strengths of using LabVIEW, which is also found in the automotive industry. A very important step in V-shape control design that is also applied in the automotive industry is the hardware-in-the-loop (HIL) test step. HIL is a technique that involves the simulation of a real system through hardware and software that allows the physical connection and testing of a real device, device under test (DUT) (e.g., electronic control units, ECU). Real signals are passed between the DUT and the simulated systems, thereby tricking the DUT into working on the final product. There are many study cases that apply the HIL technique used in in-vehicle testing system [19], battery management system testing [20], or electrical vehicle management electronics testing [21].

2.4 LabVIEW in remote control

Lately, the remote (online) laboratories became an important component of the educational process. The remote labs became a vital component during the COVID pandemic. During the last two decades, many educators implemented different laboratories controlled remotely based on different software solutions and covering different education fields. Due to the facilities of hardware controls, interoperability with other software applications, and network communications, LabVIEW represents one of these solutions. Simple to complex solutions of remote laboratories are implemented and described in the specialty literature. An online experiment for eddy current magnetic brake study based on LabVIEW and NI myRIO is presented in [22], while in Ref. [23], a Digital Oscilloscope study is presented as part of instrumentation and measurements course. Both solutions are based on a rapid but old technology called LabVIEW web publishing tool. More complex solutions were developed, such as the combination between the VISIR hardware platform and iLab remote control and management software platform offered by MIT [24]. A solution based on web programming techniques and web services under LabVIEW is presented in Ref. [25].

To increase the flexibility of web publishing of the LabVIEW developed application, the G Web Development Software was released by NI. This software package allows developing web-based user interfaces for applications developed in LabVIEW, Phyton, and/or C#, without the knowledge of classical web programming [26].

Another solution for remote control of testing and measuring instruments is the usage of the gRPC framework. This framework is an open source that was released by Google in 2015 and is based on Remote Procedure Call framework. gRPC is based on HTTP/2 and therefore is easy to integrate with a modern internet infrastructure. This framework is mainly dedicated to developing services, and there are eleven officially supported languages. There are several other languages that can support gRPC, and one of these is LabVIEW. The LabVIEW gRPC support can be found on GitHub (https://github.com/ni/grpc-labview).

3. Conclusion

Virtual instrumentation represents a strong and useful concept that is used in a large number of domains, from all engineering domains to medical or educational ones.

The combination of the modular hardware and the customizable software offers the possibility to develop simple and rapid applications but at the same time can be used for developing very complex applications applied to the most recent and modern domains.

One solution for implementing the virtual instrumentation concept is LabVIEW. Due to the graphical programming technique, this language can be rapidly learned and used by non-programming specialists. The NI LabVIEW applications can be developed by non-programming specialists due to the "programming like you think" approach and can cover programming of the RT and FPGA systems, HIL simulation systems, electrical power measurement systems based on dedicated LabVIEW toolkit, remote controlled systems based on web services and G Web Development Software, and numerous other systems and applications.

Author details

Petru Adrian Cotfas[1*], Daniel Tudor Cotfas[1] and Horia Hedesiu[2]

1 Faculty of Electrical Engineering and Computer Science, Transilvania University of Brasov, Electronics and Computers Department, Romania

2 Faculty of Electrical Engineering, Technical University of Cluj-Napoca, Electrical Machines and Drives Department, Romania

*Address all correspondence to: pcotfas@unitbv.ro

References

[1] Jennings R, De la Cueva F. LabVIEW Graphical Programming. 5th ed. New York: McGraw-Hill Education; 2020. Available from:https://www.accessengineeringlibrary.com/content/book/9781260135268

[2] Graphical system design. Available from: https://en.wikipedia.org/wiki/Graphical_system_design [Accessed: Oct. 03, 2023]

[3] Graphical system design basics: Accelerating development time and bringing embedded design to the masses. Available from: https://www.ni.com/en/shop/labview/graphical-system-design-basics--accelerating-development-time-an.html [Accessed: Oct. 03, 2023]

[4] Lukindo A. LabVIEW Queued State Machine Producer-Consumer Architecture. Canada: Mezintel Inc. Calgary AB; 2007. Available from: https://www.mezintel.com/blog/labview-queued-state-machine/ [Accessed: Oct. 03, 2023]

[5] Schwartz M, Manickum O. Programming Arduino with LabVIEW. Birmingham, UK: Packt Publishing; 2015

[6] Quang DT. LabVIEW and open embedded system [internet]. In: LabVIEW - A Flexible Environment for Modeling and Daily Laboratory Use. London: IntechOpen; 2021. DOI: 10.5772/intechopen.98271

[7] Kehtarnavaz N, Mahotra S. Digital Signal Processing Laboratory: LabVIEW-Based FPGA Implementation. Boca Raton, FL: Brown Walker Press; 2010

[8] Stratoudakis T. Introduction to LabVIEW FPGA for RF, Radar, and Electronic Warfare Applications. Norwood, MA: Artech House; 2021

[9] Ortiz A, Mendez E, Balderas D, Ponce P, Macias I, Molina A. Hardware implementation of metaheuristics through LabVIEW FPGA. Applied Soft Computing. 2021;**113**(Part A):29. DOI: 10.1016/j.asoc.2021.107908. Available form: https://0a10af1i5-y-https-doi-org.z.e-nformation.ro/

[10] Chouder A, Silvestre S, Taghezouit B, Karatepe E. Monitoring, modelling and simulation of PV systems using LabVIEW. Solar Energy. 2013;**91**:337-349. Available from: https://0a10aa2ix-y-https-doi-org.z.e-nformation.ro/10.1016/j.solener.2012.09.016

[11] Ahmed CB, Kassas M, Ahmed SE. LabVIEW based PV panel online characteristics and parameters estimation. Procedia Computer Science. 2015;**52**:876-882. DOI: 10.1016/j.procs.2015.05.145

[12] Cotfas PA, Cotfas DT. Design and implementation of RELab system to study the solar and wind energy. Measurement. 2016;**93**:94-101. DOI: 10.1016/j.measurement.2016.06.060

[13] Vukovic F, Milardic V, Milos D, Filipovic-Grcic B, Stipetic N, Franc B. Development and laboratory testing of a lightning current measurement system for wind turbines. Electric Power Systems Research. 2023;**223**:7. DOI: 10.1016/j.epsr.2023.109572

[14] Ogaili AAF, Jaber AA, Hamzah MN. Wind turbine blades fault diagnosis based on vibration dataset analysis. Data in Brief. 2023;**49**:6. DOI: 10.1016/j.dib.2023.109414

[15] Sudharsan GS, Vishnupriyan J, Anand KV, Chidambarathanu K. Wind Turbine Instrumentation System using Labview. In: Third International Conference on Intelligent Computing Instrumentation and Control Technologies (ICICICT), Kannur, India. IEEE. 2022. pp. 121-124. DOI: 10.1109/ICICICT54557.2022.9917754

[16] Kim M, Choi BY, Kang K-M, Lee H, Lee Y-S, Won C-Y. Energy monitoring system of AC/DC hybrid microgrid systems using LabVIEW. In: 23rd International Conference on Electrical Machines and Systems (ICEMS), Hamamatsu, Japan: IEEE; 2020. pp. 489-493. DOI: 10.23919/ICEMS50442.2020.9290836

[17] Aissou S, Rekioua D, Mezzai N, Rekioua T, Bacha S. Modeling and control of hybrid photovoltaic wind power system with battery storage. Energy Conversion and Management. 2015;**89**:615-625. DOI: 10.1016/j.enconman.2014.10.034

[18] Perz P, Malujda I, Wilczyński D, Tarkowski P. Methods of controlling a hybrid positioning system using LabVIEW. Procedia Engineering. 2017;**177**:339-346. DOI: 10.1016/j.proeng.2017.02.235

[19] Kumar AV. Creating a HIL simulation and in-vehicle test system using CompactRIO, FPGA, and LabVIEW, Tata Motors. Available from: https://www.ni.com/en/innovations/case-studies/19/creating-a-hil-simulation-and-in-vehicle-test-system-using-compactrio-fpga-and-labview.html [Accessed: Oct. 22, 2023]

[20] Hwang SU. Developing an HIL simulator for testing battery management system logic Sung-Up Hwang, Control Works. Available from: https://www.ni.com/en/innovations/case-studies/19/developing-an-hil-simulator-for-testing-battery-management-system-logic.html [Accessed: Oct. 22, 2023]

[21] Martínez J. Creating an Electric Vehicle HIL Test Using LabVIEW and CompactRIO Asoindel. Available from: https://www.ni.com/en/innovations/case-studies/19/creating-an-electric-vehicle-hil-test-using-labview-and-compactrio.html [Accessed: Oct. 22, 2023]

[22] Odema M, Adly I, Ghali HA. LabVIEW-based interactive remote experimentation implementation using NI myRIO. In: International Conference on Innovative Trends in Computer Engineering (ITCE), Aswan, Egypt. IEEE. 2019. pp. 214-218. DOI: 10.1109/ITCE.2019.8646602

[23] Hariton A, Zet C, Vremera E, Fosalau C. Remote laboratory – Study of digital oscilloscope, 2020. In: International Conference and Exposition on Electrical and Power Engineering (EPE), Iasi, Romania. IEEE. 2020. pp. 495-500. DOI: 10.1109/EPE50722.2020.9305551

[24] Zutin DG, Auer ME, Gustavsson I. A VISIR lab server for the iLab shared architecture. International Journal of Online Engineering. Barcelona, Spain: IATED 2016;7:14-17. DOI: 10.3991/ijoe.v7iS1.1754

[25] Cotfas PA, Cotfas DT, O.M. Machidon: Remote laboratories based on Labview web services. In: 8th International Conference on Education and New Learning Technologies - EDULEARN16, Barcelona, Spain. 2016. pp. 3683-3689. ISBN: 978-84-608-8860-4. ISSN: 2340-1117. DOI: 10.21125/edulearn.2016.1844

[26] Milan R. How to build a web UI for your LabVIEW-based test system. 2022. Available from: https://forums.ni.com/t5/Example-Code/How-to-Build-a-Web-UI-for-Your-LabVIEW-Based-Test-System/ta-p/4233446 [Accessed: Oct. 22, 2023]

Chapter 2

Using of Virtual Instrumentation in the Teaching of Autotronics

Matúš Danko, Ondrej Hock, Jozef Šedo and Branislav Hanko

Abstract

This chapter presents using of virtual instrumentation for improvement of teaching and demonstration of communication of one of the most popular automotive buses CAN BUS. Virtual instrumentation is used in several ways for easier understanding of communication by using the automotive bus. The first way is a demonstration by visualization of data of programable gateway. The second way of using virtual instrumentation is the visualization of the instrument cluster which is controlled by the student's development kit with the microprocessor. The visualization is also used for the demonstrational panel which consists of a development kit, instrument cluster, turbocharger, and accelerator pedal. The communication with these components is also visualized. The last usage of virtual instrumentation is controlling real instrument clusters from Skoda and VW vehicles using LabVIEW and CAN BUS interfaces. In both, controlling and visualizing of instrument cluster are used the same messages as in real vehicle, so it is easier for students to understand the structure of communication through automotive buses of the real vehicle.

Keywords: virtual instrumentation, LabVIEW, CAN BUS, automotive bus, autotronics

1. Introduction

The faculty of Electrical Engineering and Information Technology at the University of Žilina (UNIZA) is one of the oldest parts of the entire university (since the university's origin in 1953). For more than a half of century, this faculty is following novel trends in electrical engineering, electronics, industry, and science.

Last decade, several car manufacturers or manufacturers of car equipment and devices (e.g. KIA, Land Rover, Volkswagen, and Brose) established their factories and also R&D departments in the region of Žilina (northwest part of Slovakia). At this time, we can see decreased demand from mentioned companies for employees in this field. Department of Mechatronics and Electronics (DME) at the University of Žilina immediately started a new study program "Autotronics" to bring the specialists focused on car parts and sensors functionality, electrical devices in vehicles, supplying systems in hybrid or electrical vehicles, programming the car control units, and understanding the auxiliary infrastructure and economic aspects of car manufacturing and transportation.

As we can see in **Figure 1**, the interdisciplinary character of the entire Department of Mechatronics and Electronics is fully integrated into the Autotronics program.

Following the modern trends in education, adapting to the demands of manufacturers and the industrial environment, and also finding the educational methods during the specific situation caused by COVID-19, we would like to introduce the educational model for selected autotronics courses provided by our department DME. As we will see in the next sections, the model composed of various parts is joined into the functional complex system through the development system LabVIEW from the NI company. Even though LabVIEW was dedicated to creating virtual instruments in a practical environment, it was attractive for education from 1980s as documented in [1]. The studies [2, 3] focus on the advantages of virtual instrumentation in LabVIEW in the engineering education process and the creation of a remote laboratory. NI responded to this trend (usage of LabVIEW in the teaching process) and now we can find many articles, forums, and recommendations for teachers and students directly on NI websites [4]. Researching the internet sources, we can find several complex systems for the education of automotive programs and courses [5, 6]. But these simulators are the systems which contain only real parts of vehicles, so customization is not possible. Also, these simulators are not delivered with visualization software. In [7], the vehicle CAN BUS simulator that is programmed in C language is presented. In [8, 9], systems where CAN BUS interface is controlled and visualized by the original software of the interface or in combination with MATLAB are presented. Thus, these systems are not fully open systems, and customization is more difficult than customization of the LabVIEW application which is used for control and visualization at the same time. This advantage of easy customization led to the cooperation of teachers and students in development: the results of selected bachelor or diploma theses can be integrated to extend the existing model.

One of the most important parts of modern vehicles is the communication between many different systems used in vehicles [7, 8]. There are several automotive buses that can be divided by transmitting speed, the medium and maximum length of the bus, or devices connected to the bus. Real-time control of the internal

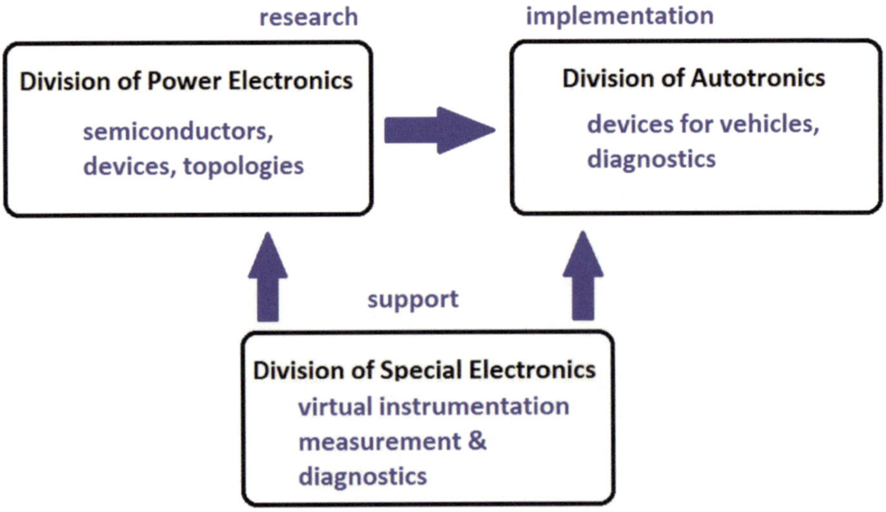

Figure 1.
Mutual cooperation between divisions of DME.

combustion engine (ICE) of the converter of the electric drive is needed high-speed communication with high reliability. For a less important system like comfort systems, the speed of the bus can be slower, and the transmission media can be simplified. Based on this division, each automotive bus is suitable for the connection of a specific group of devices. Because of its properties, the CAN BUS is used in various systems which make it one of the most common and well-known automotive buses. For purpose of teaching how this automotive bus works, the workbenches of the laboratory were modified. Every workbench contains a connector for connecting students' evaluation boards with the microprocessor and proposed systems to CAN BUS. Workbenches are connected between each other and to the programable gateway which is connected to the superior workbench. This programable gateway is one of the proposed LabVIEW applications, which is used during the teaching process. During the operation of this gateway, the structure of CAN BUS is demonstrated, which means that there are different frames, the content of each type of frame, and the processing of messages. Following LabVIEW applications are used with components from real vehicles mainly instruments cluster, because on this vehicle component it is easy to see the correctness of communication. These applications serve for demonstration of transmitting data by using scaling and offset which are necessarily high and negative values that can be transmitted directly. From selected vehicle components, demonstration panel which is used as the source of communication for the proposed LabVIEW application with visualization of the instrument cluster was created. The last proposed application generates messages/signals for controlling instrument clusters.

2. CAN BUS

The CAN (Controlled Area Network) bus was developed by BOSCH company as a multimaster messaging system. The CAN BUS specifies a maximum speed up to 1 megabit per second (1 Mbps). The CAN BUS network sends small data blocks point to point from node A to node B under the supervision of a central bus, instead of large blocks like traditional networks such as USB or Ethernet. In CAN bus, data are sent and received in many short messages like temperature, engine speed, pressure, etc. This system of many short messages ensures data consistency for every node in the system [10, 11].

CAN BUS utilizes differential transmission of signals, which results in robust noise immunity and fault tolerance. The transmission of differential signals reduces noise coupling and enables high transmission speed by using twisted pair of wires. The differential transmission utilizes current flowing in each signal wire with the same amplitude but in the opposite direction, which results in a field cancelation effect. This is key for low emission of electromagnetic noise and also for high immunity of the electromagnetic noise.

The CAN protocol defines two logic states of the bus, dominant and recessive states. ISO-11898 defines a differential voltage which represents dominant and recessive state, respectively, bits as shown in **Figure 2**. In the dominant state (a logic '0'), the differential voltage between CANL and CANH must be greater than the minimum threshold value (under 0.5 V for receiver input and under 1.5 V for transmitter output). In the recessive state (a logic "1"), the differential voltage across CANH and CANL must be greater than the minimum threshold. The dominant bit wakes up the recessive bit on the bus to achieve non-destructive bit arbitration [11, 12].

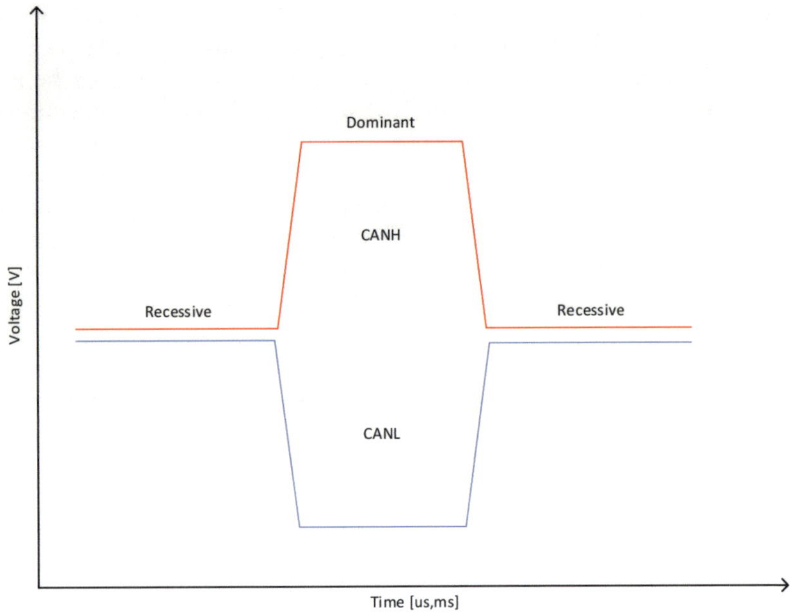

Figure 2.
Differential voltage defined by ISO-11898.

The CAN BUS is a serial communication bus defined by the International Organization for Standardization (ISO), which was developed for the automotive industry for the replacement of a complex wiring harness with a simple two-wire bus. The ISO specification requires high immunity against the electromagnetic interference and the ability to self-diagnostic and the possibility of data error corrections. These features helped to increase the popularity of CAN BUS in various fields of industries like manufacturing, building automation, medicine, etc. The communication protocol CAN BUS by ISO-11898:2003 describes how information is transferred between network devices and corresponds to the OSI (Open Systems Interconnection) model. This model is defined in terms of layers. The physical layer defines communication between nodes that are connected by physical media. The two lowest layers of the seven-layer OSI/ISO model which are the physical and data-link layers are defined by ISO 11898 [11, 13].

Figure 3 shows the application layer that establishes a communication link with a specific higher-level application protocol, such as the vendor-independent CANopen protocol. This protocol is supported by an international group of users and manufacturers from all over the world. Many protocols are designed for the specific applications such as industrial automation, diesel engines, or aerospace [9]. Other examples of CAN-based industry standard protocols are CAN KVASER and CAN from Rockwell Automation's DeviceNet [13, 14]. Even though ISO 11898 defines the electrical specification of wires and connector, it does not define wires and connector directly. ISO 11898 defines the requirement for termination resistors at each end of the bus with a value of 120 Ω, which can be seen in **Figure 4**. Most often, this termination resistor is part of hardware dedicated to CAN BUS, so a simple connection of two devices is necessary for the establishment of the communication. The disadvantage of the fact that most devices contain termination resistors is that, if lots of devices are used, the resistance of the bus will be too low. In this case, the termination resistor of some devices must be removed [12, 15].

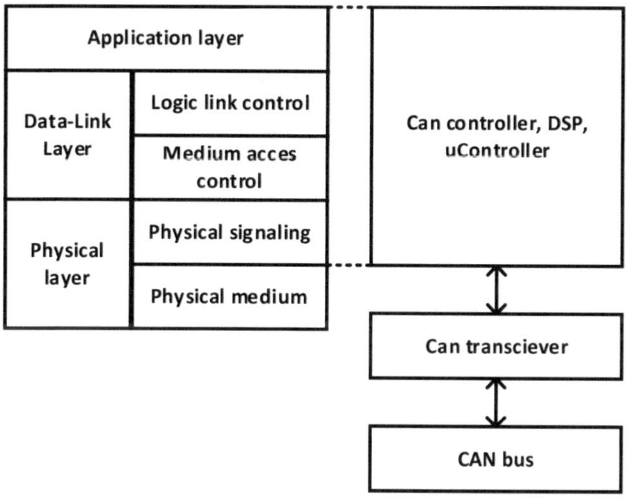

Figure 3.
Physical and data-link layer definition by ISO 11898 Standard Architecture.

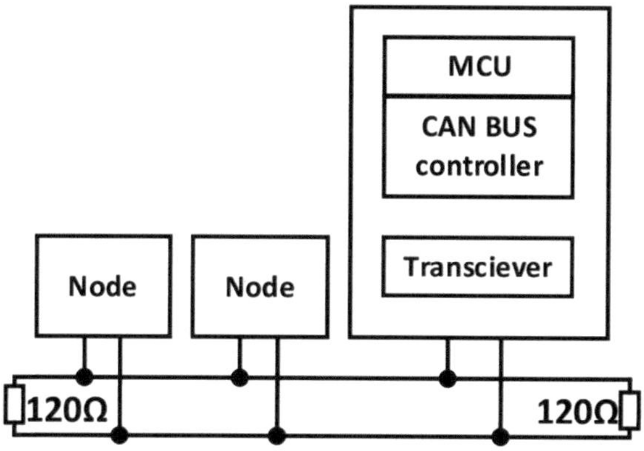

Figure 4.
Connection of nodes to bus with termination resistors.

The CAN communication protocol is a multiple access protocol in terms of the carrier signal with the detection of collision and the message priority decision (CSMA/CD + AMP). CSMA means that each bus node must wait a defined time of inactivity before attempting to send a message. CD + AMP means that collisions of the transmission of more nodes are solved using bit arbitration based on the priority of each message. This priority is defined in the identifier field of each message. The access to the bus and transmission is always given by the identifier field with the higher priority [11, 14]. The communication via CAN BUS is realized with a different type of message or frame with two lengths of identifier. The identifier can be standard with a length of 11-bit or extended with a 29-bit length. The first is the data frame which is used for the transmission of data between nodes of the bus. The second type of frame is the remote frame which is transmitted to the bus for the request of data transmission of a specific node. The third type of frame is the error frame, which is transmitted by the node, which detects the error. The last type of frame is the overload frame which is transmitted by a node to delay the transmission data of the bus to get extra time to process data.

Transfer speed defined by the ISO-11898 standard are from 125 kbps to 1 Mbps. This standard defines standard 11-bit identifier an extended 29-bit identifier [9, 16]. The content of standard 11-bit identifier is shown in **Figure 5**. This identifier provides up to 2048 different message identifiers, while the extended 29-bit identifier, which is shown in **Figure 6** provides up to 537 million identifiers [11, 17].

The meaning of the bit fields in **Figure 5** is as follows:

- SOF – one dominant Start of Frame (SOF) bit indicates the start of a message and is also synchronized bus nodes after being idle.

- Identifier – the standard 11-bit CAN identifier determines the priority of the message, the lower value in binary form means a higher priority.

- RTR – the single remote transfer request (RTR) bit is dominant when information is requested from another node. All nodes of the bus receive this request, but the identifier defined the specified node of the bus. Corresponding data are also received by all nodes and used by all involved nodes. In this way, all data used in the system are uniform.

- IDE – the single Identifier Dominant Extension (IDE) bit means that a standard CAN identifier is transmitted without an extension.

- r0 – reserved bit (for possible of future supplementation of standard).

- DLC – the 4-bit field of Data Length Code (DLC) defined the number of data bytes to be transmitted.

- Data – up to 64 data bits can be transmitted.

- CRC –15 bits with delimiter used for Cyclic Redundancy Check (CRC) contains a checksum of previous application data for error detection.

- ACK – each node receiving an accurate message overwrites this recessive bit in the original message with a dominant bit, indicating that an error-free message has been sent. If the receiving node detects an error and leaves this bit recessive, it discards the message, and the sending node retries the message after rebitting. In this way, each node acknowledges (ACK) the integrity of its data. ACK is 2 bits, one is for the acknowledgment bit and the other is used as the delimiter.

SOF	11-bit identifier	RTR	IDE	r0	DLC	8 bytes data	CRC	ACK	EOF	IFS

Figure 5.
Standard CAN BUS frame with an 11-bit identifier.

SOF	11-bit identifier	SRR	IDE	18-bit identifier	RTR	r0	r1	DLC	8 bytes data	CRC	ACK	EOF	IFS

Figure 6.
Extended CAN bus frame with a 29-bit identifier.

- EOF – this 7-bit end-of-frame (EOF) field marks the end of a CAN (message) frame and disables bit padding, indicating a padding error when dominant. When 5 bits of the same logic level occur in a row during normal operation, a bit of the opposite logic level is inserted into the data.
- IFS - this 7-bit inter-frame space (IFS) contains the time it takes for the controller to move a correctly received frame to its correct position in the message buffer area.

As shown in **Figure 6**, the Extended CAN message is the same as the standard message with the addition of:

- SRR – the Surrogate Remote Request (SRR) bit replaces the RTR bit in the standard message location as an extended format placeholder.
- IDE – a recessive bit in the identifier extension (IDE) indicates that more identifier bits follow. The 18-bit extension follows the IDE.
- r1 – After the RTR and r0 bits, an additional reserve bit was included before the DLC bit.

3. Analysis and processing of CAN BUS messages using programable gateway

Data transfer of CAN BUS using fully programable gateway is demonstrated. During the operation of this gateway, the structure of CAN BUS communication is demonstrated, which means that there are different frames, the content of each type of frame, and the processing of messages. A gateway, in general, is a device that is used for the connection of two different buses with the different transmission protocols or buses with the same protocol but with different speeds. A gateway is used for two main reasons: the first reason is the length of the bus at a maximum speed of 1 mbit/s which is only 40 m. With the existing arrangement, it is not possible to connect all students' workbenches with a superior workbench or teacher's workbench with one bus. The second reason is the connection of this superior workbench which is used for remote access to students' workbenches. By using the gateway, it is possible to connect these two CAN BUSES which can use different speeds and is possible to filter messages and resend them from the first bus to the other and vice versa. This gateway can be controlled from a superior workbench with LabVIEW. From this workbench, the message of both parts of the laboratory CAN BUS is visualized. Also, the filtering of messages is controlled from this workbench. This gateway, which block diagram is shown in **Figure 7**, utilizes NI CompactRIO 9082 as the main part. It is the system that combines a real-time processor for processing and monitoring and FPGA for high-speed logic and precise timing [18]. This system can use up to eight modules; however, this gateway utilizes only two modules. The first of the modules is CAN BUS module NI9853 which disposes of with two high-speed ports with standard DB9 connectors. The second module is the module for SD card NI9802 which is used for transferring data logs to SD cards.

As mentioned before, the system CompactRIO combines real-time operation using a processor and FPGA; therefore, the programming approach is different from

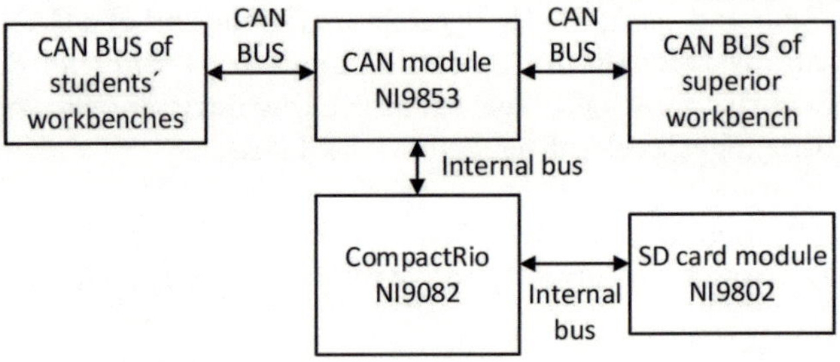

Figure 7.
Block diagram of the connection of gateway.

programming other NI measuring and communication cards, NI hardware, or third-party hardware. LabVIEW project is divided into three programs, each for one level of processing – FPGA, real-time, and host computer as we can see in **Figure 8**. The main program for the acquisition or transmission of data (signals) is based on the usage of specific modules, because some modules are dedicated for usage with FPGA processing and some modules for processing with the processor of CompactRIO. Both used modules, for CAN BUS communication and data transmission to SD cards, are dedicated to the FPGA level. At this level, the FPGA program is used for the acquisition and transmission data of CAN modules like identifier, type of frame, DLC, and an array of data. This main FPGA program is also used for saving data logs to SD cards. This program use mainly functions from the FPGA library. The next level is a real-time program that can be used for processing and visualization, but it is limited in the computation power of the CompactRIO processor. Instead of processing, this level is used for update of shared variables which are used for transmitting data to the host computer. In the host computer data, processing and visualization are limited by the computation power of the PC that is used. This approach is suitable where for visualization a large number of indicators, graphs, etc., like virtual instrument

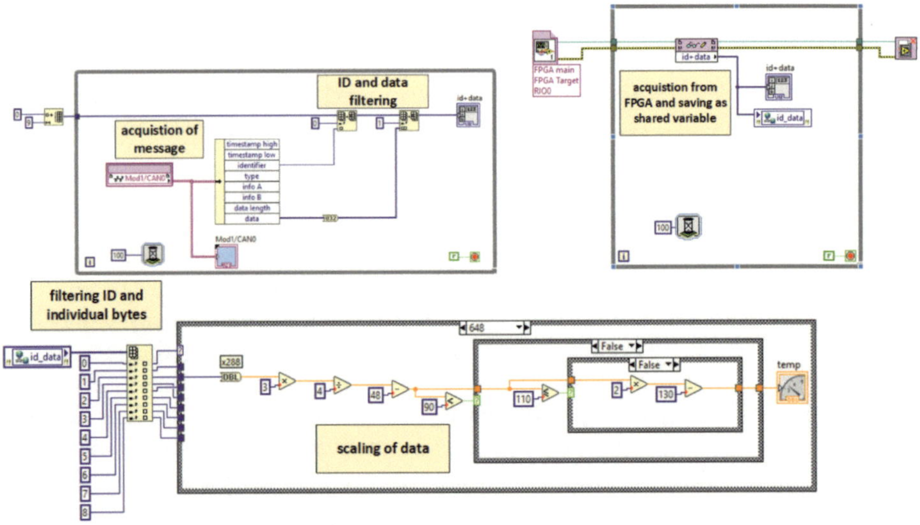

Figure 8.
Block diagram of three programs, each for one level of message processing, top left for FPGA, top right for real time, and bottom for the host computer.

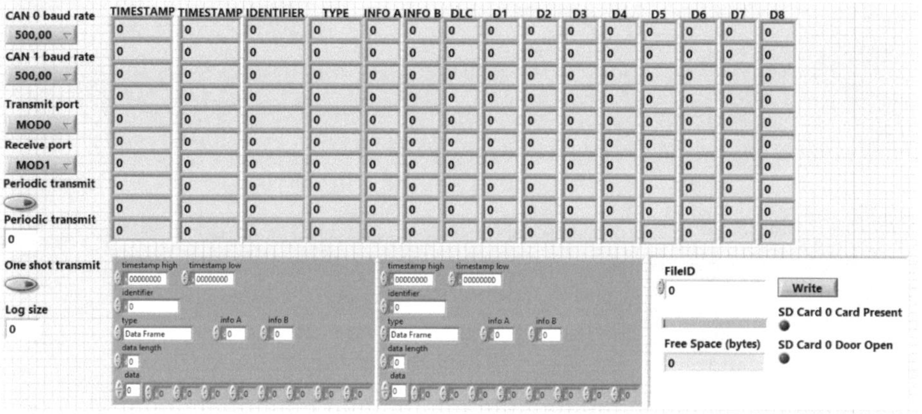

Figure 9.
Front panel of the program for control of the gateway.

clusters application were used. Also, the complex calculation is easier with this approach because FPGA uses only fixed-point number, so some functions of libraries are missing. In this last level, data from the first or second bus are visualized.

Using front panel controls, it is possible to control the gateway as shown in **Figure 9**, a basic setting like the speed of each port and the way how the gateway works. The default mode of the gateway is resending data from the first to the second bus without filtering, so the gateway works as a repeater. This mode work with buses with the same speed or resending data from the bus with a lower speed to a bus with a higher speed. To keep the gateway simple, no buffer is needed. In this mode, data are resending when are available on the bus. The next mode is when the gateway sends data on the bus periodically and data are updated when data on the first bus were changed. This mode is used for communication with real hardware because this hardware needs precise periodic communication. The control unit controls the time between messages, and if this time is smaller or greater than the defined time control unit evaluates it as an error in the communication. From the front panel, the size of the log and the start of logging are controlled. The last mode of the operation of the gateway is the filtering work with the same speed of bus or from lower speed bus to higher speed bus because of buffering. This mode of operation has no preset filtering, and filtering is realized in the block diagram individually depending on devices connected to buses.

4. Demonstration panel of CAN BUS communication of control unit with instrument cluster and VNT turbocharger actuators and its visualization

For the demonstration of real-time communication of the control unit with sensor and actuator, a demonstration panel was created. This panel with visualization is shown in **Figure 10**. This demonstrational panel consists of a development kit with a processor with a CAN BUS module which simulated some messages between the control unit of the internal combustion engine and some sensors and actuators based on the position of the accelerator pedal. In **Figure 11**, we can see block diagram with connection of these components. This communication demonstrates the communication

Figure 10.
Demonstration panel with the accelerator pedal, simulated control unit, instrument cluster, and turbochargers.

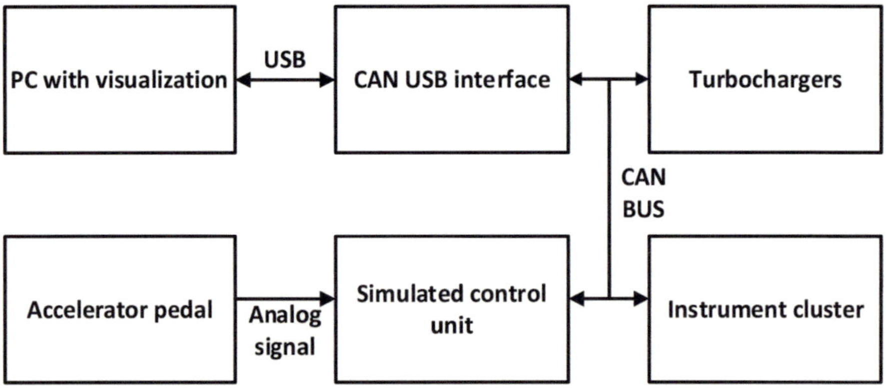

Figure 11.
Block diagram of demonstration panel and its visualization.

of real components which used scaling and offset for transmission of negative values and values greater than 255, which is the maximum value for the transmission.

The accelerator pedal is connected to two channels of development kit ADC. Two channels are used in the vehicle because of safety reasons. Both signals from the accelerator position sensor are voltage signals in the range from 0 to +5VDC and are equivalent to the position of the pedal. Most often one signal is the half-value of the other signal, or signals have opposite values. This secures that it is possible to check the validity of the signal from the accelerator pedal position sensor. If signals are not in the right relation, the signal is not valid, and the control unit works in safe mode. The next parts of the demonstration panel are two VNT turbochargers. These turbochargers use vanes to regulate boost pressure. These vanes are powered by the electric actuator, and in modern vehicles, actuators are controlled directly by CAN BUS. The last part of this demonstrational panel is the instrument cluster which shows engine speed, coolant temperature and speed of the vehicle, and some indicators. Some signals like fuel-level indicators and some warning indicators are still analog in vehicles.

For communications PC with demonstration panel USB-CAN interface, Kvaser Leaf semi-pro was chosen because of the support of the driver for LabVIEW [19]. LabVIEW drivers contain a library with functions for programming applications with this interface. Function of this library and math library are used most often in this application. This interface is capable of having speed up to 1 mbit/s and supports standard and also extended data frames. For the measurement of analog signals like accelerator pedal position, myRIO was used. For visualization, which can be seen in **Figure 12**, a virtual instrument cluster was created using gauges and some decorative elements as much as a real instrument cluster.

The virtual instrument clusters show engine speed, vehicle speed, coolant temperature, and odometer as a trip which is calculated from vehicle speed. As mentioned before, fuel level is an analog value and on the demonstration panel, the fuel level sensor is not used (**Figure 13**).

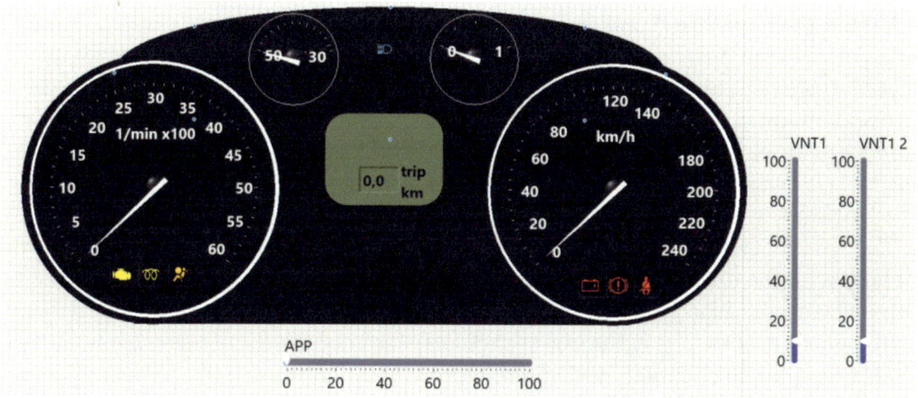

Figure 12.
Visualization of demonstration panel as virtual instrument cluster and positions of the accelerator pedal and turbochargers actuators.

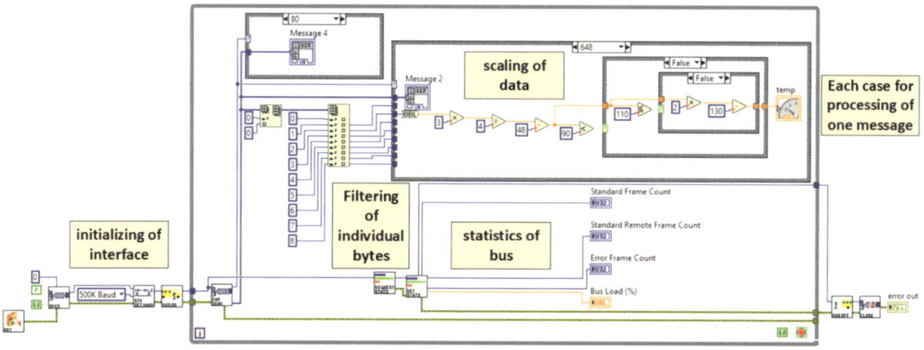

Figure 13.
Block diagram of visualization of demonstration panel as virtual instrument cluster and positions of the accelerator pedal and turbochargers actuators.

As previously mentioned, visualization must process data as a real instrument cluster (**Figure 14**). The values of engine speed and vehicle speed are processed with offset and scale, similarly, as shown in **Figure 15**. The processing of coolant temperature is respective value on indicator that is different from other indicators. It is caused by hysteresis of the thermostat of ICE cooling, so during operation the temperature rises and falls repeatedly. This means that without modifying of processing temperature, indicator will show the temperature the same way. Temperature is processed to indicator to show 90°C in the operating range of temperature (85–110°C) as shown in **Figure 13**. This means that up to operating temperature, indicator shows real temperature and in the interval of operating temperature shows 90°C. It is a temperature higher than the operating temperature, and the temperature on the indicator rises with a greater slope.

At the task of creating messages, students learn about scale and offset in the calculation of data through CAN BUS. Scaling and offset are necessary because of the structure of the data protocol of CAN BUS. In the description of the protocol was mentioned that it is possible to transmit only unsigned 8-bit numbers. That means to transmit negative values, like temperatures, usage of offset is required. For transmission of a number higher than 255, which is a maximum of one byte, scaling is needed. The second way of transmission of high values is like the speed of the engine, and the 16-bit number is split into two 8-bit numbers and these numbers are transmitted to the bus.

Figure 14 shows the structure of the real message where we can see the usage of two bytes for transmission and usage offset and scale. This message is defined by Fleet Management System (FMS) which is the protocol used in trucks of the main European manufacturers like Scania, Man, Volvo, etc. [20]. This protocol is used as an example because it is one of the few publicly available protocols. Almost all manufacturers of passenger vehicles, respectively, concern of vehicles manufacturer use their own protocols which are not usually publicly available.

This virtual instrument cluster can be used also with students' development kits not only with the demonstration panel. It is easier for students to debug their program with real communication using this visualization instead of using a real instrument cluster which is quite big in the workspace and does not need a power supply. When students control this virtual instrument cluster, they must use defined messages to control the tachometer, speedometer coolant gauge, and other messages, for example warning indicators are defined by our visualization.

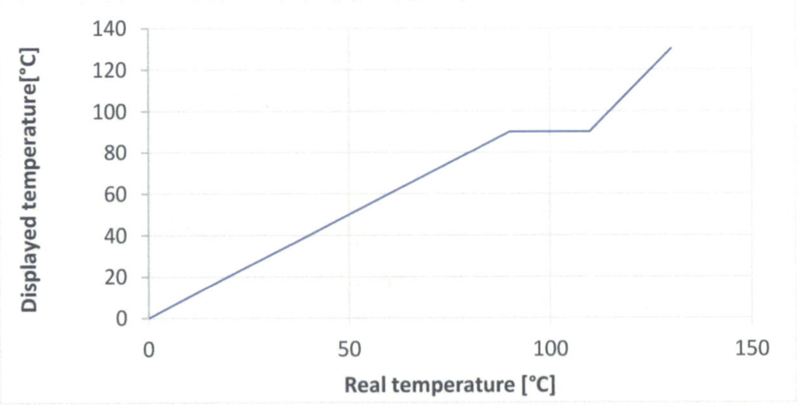

Figure 14.
Nonlinearized curve of temperature indicator.

Identifier	Tmin	Tmax	PGN	Default Priority	Message Type	DPext	DP	Group Extension	Source	Destination	Standard/ proprietary
18F00503	100	100	00F005	6	Cyclic	0	0	05	03		Std

Sender	Receiver	Byte	Bit	Length	Explanation	State	Resolution	Offset	Operating range Data range	Unit	Note
	X	1	1	8	**SelectedGear (Selected Gear, - rev, + forw, 0 neut, 126 park)**		1	-125	-125 - 125		
	X				Park	0xFB					
	X				Error	0xFE					
	X				NotAvailable	0xFF					
		2	1	16	**ActualGearRatio (Actual Gear Ratio)**		0.001	0	0 - 64.255		
					Error	0xFE00					
					NotAvailable	0xFF00					
	X	4	1	8	**CurrentGear (Current Gear, - rev, + forw, 0 neut, 126 park)**		1	-125	-125 - 125	Gear	
	X				Park	0xFB					
	X				Error	0xFE					
	X				NotAvailable	0xFF					
		5	1	16	**TransmissionRqstedRange (Transmission Requested Range)**		1	0	0 - 255	Ascii	1)
					Error	0x00					
					D	0x2044					
					N	0x204E					
					R	0x2052					
					NotAvailable	0xFFFF					
		7	1	16	**TransmissionCurrentRange (Transmission Current Range)**		1	0	0 - 255	Ascii	2)
					Error	0x00					
					D	0x2044					
					N	0x204E					
					R	0x2052					
					NotAvailable	0xFFFF					

Figure 15.
Structure of real CAN BUS message [12].

5. Control of instrument clusters using LabVIEW

Visualization of instrument clusters can be modified to be used for controlling real instrument clusters. Gauges are used as controls, so the value is chosen with the position of the gauge needle. Based on the position of the needle of gauges, data of CAN BUS messages are calculated. Data are calculated opposite as for visualization, for example for visualization of engine speed, data from the bus are divided by 4, and for control, data are multiplied by 4. Block diagram for calculation of data with visualization is shown in **Figure 16**. For demonstration and easier understanding of calculation of message data, subresults and results are shown in decimal and also hexadecimal form, as shown in **Figure 17**. For communication, CAN-USB interface Kvaser Leaf semi-pro is used. This control of the instrument cluster by using virtual instrumentation serves for the analysis of communication with the real instrument cluster. With this analysis, it is easier for the student to program their own development kit for controlling the real instrument cluster or its visualization.

For the demonstration of the advantages and disadvantages of using the bus in the vehicle, an instrument cluster from an old BMW was used, which was constructed without any automotive bus. Processing signals are shown in **Figure 18**.

On this BMW cluster, all indicators and gauges are controlled by an analog signal. The front panel of the program for control of which cluster is shown in **Figure 19**.

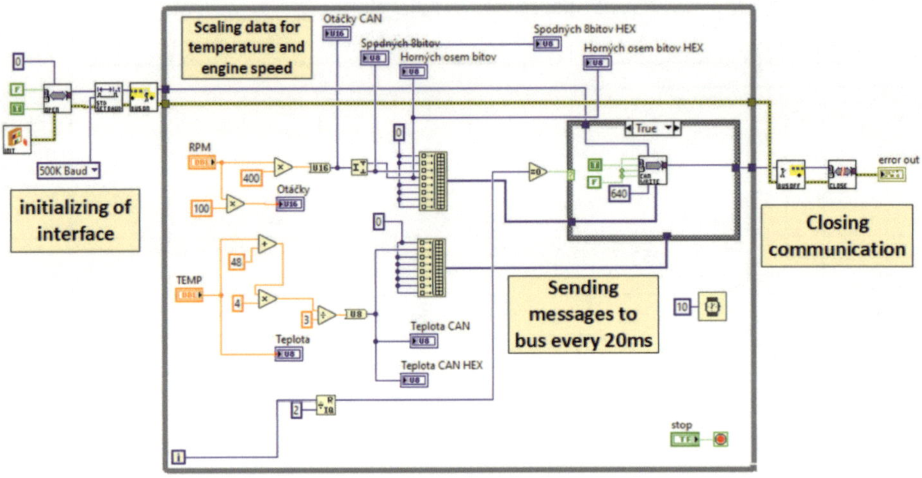

Figure 16.
Block diagram of LabVIEW program for control of VW instrument cluster through visualization with the calculation of values in decimal and hexadecimal format.

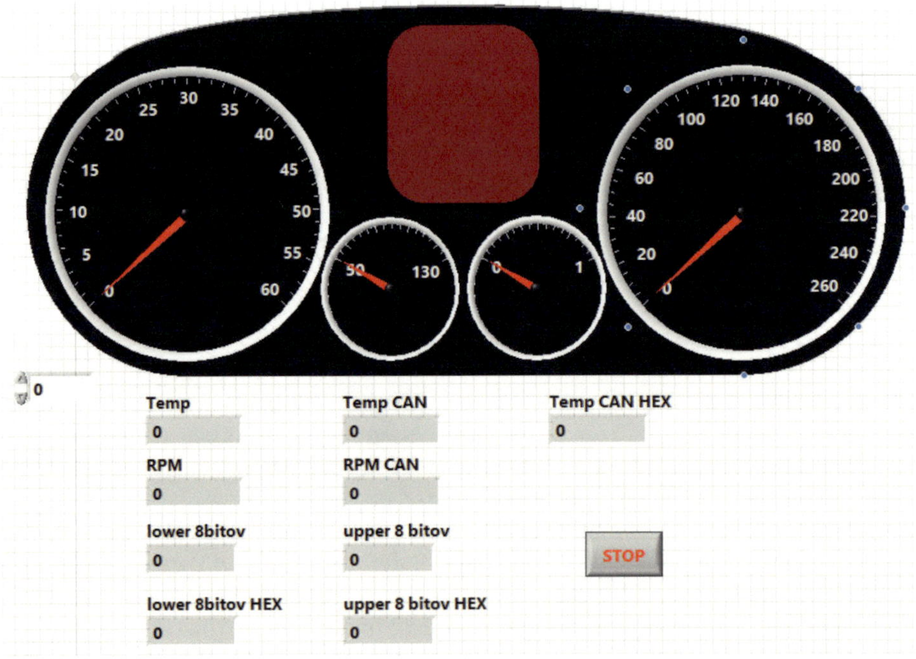

Figure 17.
Control of VW instrument cluster through visualization with the calculation of values in decimal and hexadecimal format.

For indicator, it means control with connection +12VDC or GND. For engine speed and vehicle speed, the hall sensor or the inductive sensor is used within the circuit for the signal processor, so the signal is a square wave with a duty cycle of 50%. NTC thermistor serves as the coolant temperature sensor, so the output signal is resistance. The potentiometer is used as a fuel-level sensor, so the output is resistance as well. These two gauges are controlled by PWM with fixed frequency and variable duty cycle.

DOI: http://dx.doi.org/10.5772/intechopen.109077

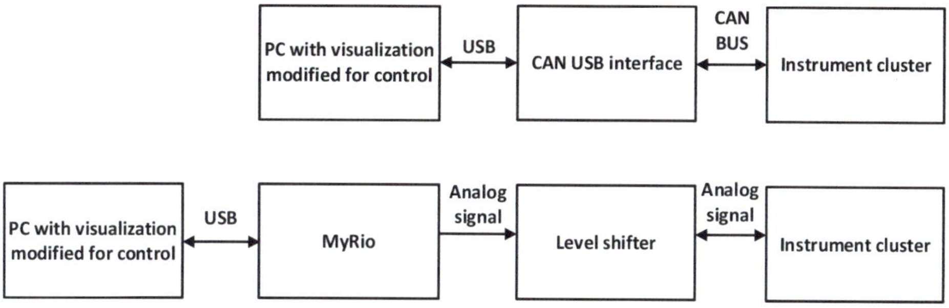

Figure 18.
Block diagram of the connection of the control instrument cluster with CAN BUS (top) and using analog signals (bottom).

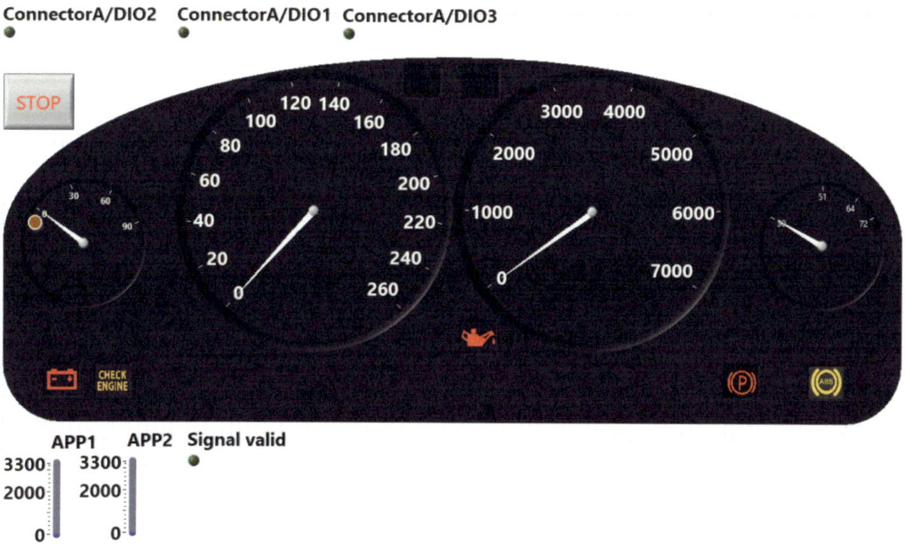

Figure 19.
Control of BMW instrument cluster through visualization with the generation of analog signals.

For generating PWM signal and logic levels for indicator, myRIO was used. MyRIO is a portable device dedicated for students to the design of control, robotics, and mechatronics system. MyRio contains two identical 34-pin connectors, connectors A and B, with 16 digital inputs/outputs, three 0–5 V analog inputs, and one analog output, 3.3 V and 5 V power output. Two voltage levels are available, because this compact device can use as digital input logic of both voltage levels, but the output has 3.3 V voltage only. Digital inputs/outputs have internal pullup resistors to 3.3 V, and this means that the state of this input without a signal will be always logic 1. Thus, these inputs are controlled by logic 0 (connection to ground). This device also contains a third 20-pin connector C with two ±10 V differential analog inputs and two outputs, eight digital input/outputs, 5 V, +15 V/ -15 V power outputs. Digital inputs/outputs of this connector have internal pulldown resistors to ground, and this means that the state of this input without a signal will be always logic 0. Thus, these inputs are controlled by logic 1 (connection to power output). Each of the analog outputs

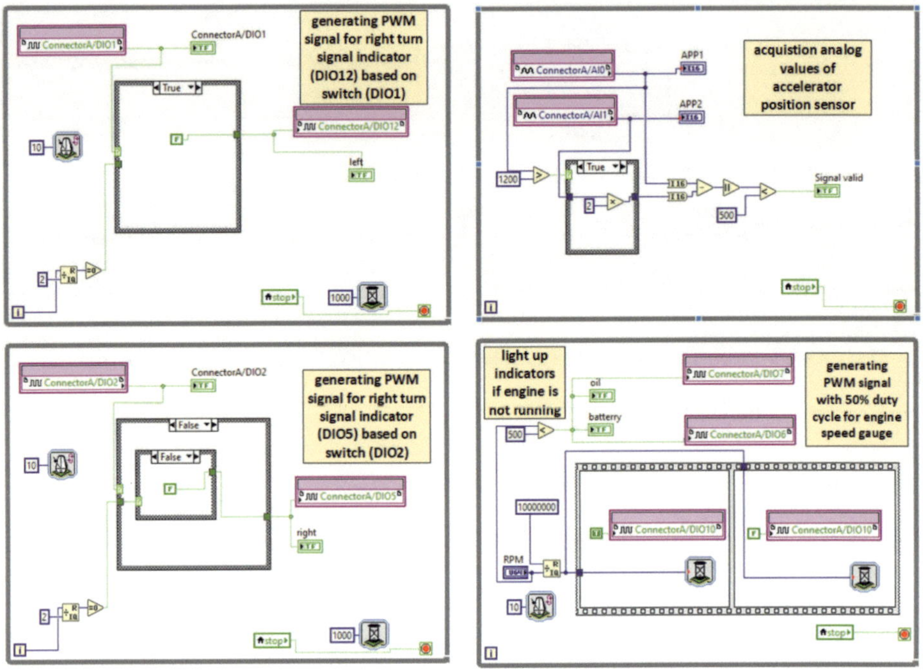

Figure 20.
Block diagram of myRIO FPGA application for generating PWM for control of BMW instrument cluster.

of all three connectors has its own digital-to-analog converter (DAC), so all analog outputs can be updated simultaneously. All digital-to-analog converters of analog outputs are controlled from FPGA by two serial buses, one for connectors A and B and the second for connector C [21]. This hardware was chosen because of the need to generate four PWM channels and eight GPIO pins in a compact package. With myRIO is used a small board with logic-level transistors to shift logic levels from 3.3 V (myRIO) to +12 V (instrument cluster). Since, myRIO contains FPGA and a processor like cRIO, the programming approach is similar. In **Figure 20**, we can see part of the block diagram of the program for generating PWM signals. For communication with CAN BUS, Kvaser Leaf semi-pro interface was used, as in other applications. Here, students may realize how CAN BUS can simplify communication and save cabling. For controlling this instrument cluster, theoretically, one message will be enough but with analog control, 22 analog signals (we use only 8 indicators and 12 signals total) would be needed.

6. Conclusions

This chapter is focused on the usage of virtual instrumentation, respectively, LabVIEW to make understanding of autotronics easier for students. Due to the growing amount of electronics in modern cars, where control units, sensors, and actuators communicate with buses, we aim for automotive bus CAN BUS. This bus was chosen because this is one of the most popular automotive buses which is now also used in various fields of industry. A demonstration panel and several applications were created to demonstrate the structure and possibilities of the communication, advantages, and disadvantage of using the bus. The first application is the programable gateway which is implemented in CompactRio with a CAN BUS module and SD card module. This programable gateway can work in different modes. In all working modes, data are transferred from the bus with a lower speed to the bus with a higher speed, or bus with the same speed because of using no buffers. In the first mode, the gateway is only resending CAN BUS frames from one bus to another like a repeater. The data are transferred only when are data available on the bus. In the next mode, data are periodically sent on the second bus, and data are refreshed when data appear on the first bus. This mode is more suitable for usage with real hardware, because real hardware measures the time between messages, to detect the error in communication. That means if the time between messages is greater or smaller than defined, the control unit evaluates corrupted communication. The last mode is the mode with filtering where filtering is implemented in a block diagram based on connected hardware, and there are no preset filters. Settings of the gateway are on the front panel of the application, like the speed of both ports, the working mode of the gateway, and the start of the log to the SD card. The students working with this gateway can better understand the frames of CAN BUS protocol, the structure of frames, etc. The second application is the visualization of the demonstration panel which consists of a development kit (control unit), accelerator pedal, and two turbochargers. The development kit simulates the control unit and based on the position of the accelerator pedal sends messages to actuators of turbocharges and to the instrument cluster. The visualization of this panel visualizes the instrument cluster, the position of actuators of both turbochargers, and the position of the accelerator pedal. This visualization can be used with students' development kits; therefore, the debugging is easier, and no additional hardware is needed. The CAN BUS communication needs scaling and offset to transmit negative values and values greater than 255, which is a maximum of one byte. With this visualization, students adopt the calculation of data for messages to control the instrument cluster. The last application controls the instrument cluster from the vehicle which instrument cluster communicates with CAN BUS and the second instrument cluster which needed analog signals. Students can see the difference in complexity of wiring with analog signals compared to simple wiring of CAN BUS.

Acknowledgements

This publication was realized with the support of Operational Program Integrated Infrastructure 2014–2020 of the project: Innovative Solutions for Propulsion, Power and Safety Components of Transport Vehicles, code ITMS 313011 V334, co-financed by the European Regional Development Fund and grant no. VEGA 1/0593/20: Research on power flow control in the network using a smart transformer.

Conflict of interest

The authors declare no conflict of interest.

Author details

Matúš Danko*, Ondrej Hock, Jozef Šedo and Branislav Hanko
Faculty of Electrical Engineering and Information Technology, Department of Mechatronics and Electronics, University of Zilina, Zilina, Slovakia

*Address all correspondence to: matus.danko@feit.uniza.sk

References

[1] Vento JA. Application of Lab VIEW in higher education laboratories. Proceedings Frontiers in Education Conference. 1988;**1988**:444-447. DOI: 10.1109/FIE.1988.35023

[2] Stefanovic M, Cvijetkovic V, Matijević M, Simic V. A lab VIEW-based remote laboratory experiments for control engineering education. Computer Applications in Engineering Education. 2011;**19**:538-549. DOI: 10.1002/cae.20334

[3] Ponce Cruz P, Molina Gutiérrez A. Lab VIEW for intelligent control research and education. In: 2010 4th IEEE International Conference on E-Learning in Industrial Electronics. Glendale, AZ, USA. 2010. pp. 47-54

[4] NI, How Do I Use LabVIEW to Teach Engineering Students?, Available online: https://www.ni.com/cs-cz/shop/labview/how-do-i-use-labview-to-teach-engineering-students.html (Accessed 26 August 2022)

[5] Automotive Training Equipment Car Chassis System Training Simulator Vocation Educational Equipment for School, Available online: https://www.alibaba.com/product-detail/Training-Simulator-Car-Training-Simulator-Automotive_1600282573224.html?spm=a2700.7724857.0.0.59bd2e76RoWeK9&s=p (Accessed 26 August 2022)

[6] Automotive Training Equipment Car Electronic Cruise Control System Trainer School Laboratory Equipment, Available online: https://www.alibaba.com/product-detail/Automotive-Training-Equipment-Car-Electronic-Cruise_1600267487948.html?spm=a2700.details.0.0.63fa69daHuHV12 (Accessed 26 August 2022)

[7] Vdovic H, Babic J, Podobnik V. Specialized vehicle CAN Bus simulator: From modelling to validation. In: 2020 5th International Conference on Smart and Sustainable Technologies (SpliTech). Split, Croatia. 2020. pp. 1-7

[8] Li Y, Liyuan T, Xiaochuan T. Simulation for vehicle comfort system based on CAN bus. In: 2010 International Conference on Computer, Mechatronics, Control and Electronic Engineering. Changchun, China. 2010. pp. 227-230

[9] Liu C, Luo F. A co-simulation-and-test method for CAN Bus system. Journal of Communications. 2013;**8**(10):681-689

[10] NXP, Inc., Bosch Controller Area Network (CAN) Version 2.0, Available online: https://www.nxp.com/docs/en/reference-manual/BCANPSV2.pdf (Accessed 26 August 2022)

[11] Texas isntrument, Introduction to the Controller Area Network (CAN). 2008, Available online: https://www.nxp.com/docs/en/reference-manual/BCANPSV2.pdf (Accessed 26 August 2022)

[12] CAN protocol used in the STM32 bootloader, Application note, Available online: file:///C:/Users/User/Downloads/an3154-can-protocol-used-in-the-stm32-bootloader-stmicroelectronics.pdf (Accessed 26 August 2022)

[13] Richards P. A CAN Physical Layer Discussion. U.S.A.: Microchip Technology Inc; 2002

[14] Watterson C. Controller Area Network (CAN) Implementation Guide, APPLICATION NOTE, Available online: https://www.analog.com/media/en/technical-documentation/

application-notes/an-1123.pdf (Accessed 26 August 2022)

[15] Connecting C166 and C500 Microcontroller to CAN, Application Note,V1.0. 2004. Available online: https://www.infineon.com/dgdl/p2900010_C166_C500_CAN.pdf?fileId=db3a304412b407950112b409fb420402 (Accessed 26 August 2022)

[16] Yang H-R, Jeong C-S, Kim H-S, Yang S-Y. A study on overall vehicle monitoring system for black box using LabVIEW. In: 2011 11th International Conference on Control, Automation and Systems. Gyeonggi-do, Korea (South). 2011. pp. 1217-1220

[17] Luo F, Li R. LIN network simulation system based On LabVIEW. In: 2010 International Conference On Computer Design and Applications. Qinhuangdao, China. 2010. pp. V5-299-V5-303

[18] cRIO-9082 Features, Available online: https://www.ni.com/docs/en-US/bundle/crio-9082-feature/page/um-purpose.html (Accessed 26 August 2022)

[19] KVASER LEAF SEMIPRO HS manual, Available online: https://canlandbucket.s3-eu-west-1.amazonaws.com/productionResourcesFiles/f5b04eef-2c58-4bab-9db6-b361d8b329fc/73-30130-00242-5_EN.pdf (Accessed 26 August 2022)

[20] CAN Communication Specification for FMS, Available online: https://bodybuilder.scania.com/content/dam/bodybuilder/local-page/thailand/CAN_Communication_Specification_for_FMS.PDF (Accessed 26 August 2022)

[21] myRIO Toolkit, Available online: https://www.ni.com/docs/en-US/bundle/labview-2019-myrio-toolkit/page/myriohelp/myriohelp.html (Accessed 26 August 2022)

Chapter 3

Virtual Instrument for Single-Phase Control of Three-Phase Reactive Power Capacitive Compensators

Alexandru Băloi, Florin Molnar-Matei and Felicia Băloi

Abstract

This chapter proposes the use of LabVIEW in a very common application both in the field of power engineering technical education and in the industrial environment. It is about compensating the reactive power with the aim of increasing the power factor. Capacitive reactive power compensators for this purpose are provided with compensation steps. The connection and disconnection of these steps according to the load curve of the consumer and the power factor at which it operates are done by means of programmable controllers. Most capacitive compensators are designed to be connected balanced to the three-phase network, in delta connection. In the situation when the reactive capacitive compensators are designed for both power factor improvement and load balancing, single-phase connection of capacitor steps is needed. The chapter proposes a LabVIEW virtual instrument that, by means of National Instruments equipment, is designed for operating condition monitoring and single-phase command. The virtual instrument was tested in the laboratory on a model consisting of a RL consumer with step-adjustable inductive component and a step-adjustable capacitive compensator. The National Instruments hardware equipment used, the modeled consumer, and its related capacitive compensator are presented. Special attention, with details of realization, is given to the virtual instrument architecture.

Keywords: reactive power compensation, shunt capacitive compensator, data acquisition, virtual instrument, single-phase command

1. Introduction

In the field of power engineering, one of the methods of increasing the efficiency of transmission and distribution facilities, and implicitly of preserving energy resources, is the concern of improving the power factor. Having a high power factor in the energy system avoids the reactive power flow from power plants to consumers through lines, stations, and substations, reducing active energy losses to the minimum level of own technological consumption.

Power factor correction is generally done by shunt capacitive compensation made by stepped capacitor banks, which, together with their switching devices (contactors, relays, static switching devices) and the automatic control, form the power reactive capacitive compensators [1]. The automatic control is done in most of the cases balanced on the three phases of the network by programmable logic controllers, the purpose of the compensation being the power factor improvement and the voltage regulation.

Another problem that leads to the reduction of the efficiency of the operation of electrical networks is their unbalanced regime. An unbalanced three-phase capacitive compensator can be used both for power factor compensation and for load balancing of unbalanced networks [2]. To be able to operate in this way, a single-phase command of the capacitive compensator is needed. A combined compensation strategy based on instantaneous reactive power theory for reactive power compensation and load balancing is also presented in Ref. [3]. Another category of devices, which, in addition to compensating the reactive power, also acts on the unbalances in the network, is the Static Var Compensator (SVC) type equipment [4–10]. They use power electronics components to control passive reactive circuit elements.

In addition to operating at a low power factor and under unbalanced regimes, the electrical networks also present other problems regarding the power quality such as harmonic regime, voltage dips, and flicker. A unitary solution for solving these problems is the installation of STATCOM type equipment [11–13].

In Ref. [14] a comparison is presented regarding the application of the solutions described above with the aim of choosing the right reactive power compensation solution depending on the particular situations in practice.

The LabVIEW environment [15] allows the monitoring and analysis of operating regimes, respectively, of the various phenomena that occur in the field of electrical networks [16, 17]. Thus, in Ref. [16] a solution for monitoring the electrical parameters of a single-phase system is presented, where the recorded data are written in files. In Ref. [17] a virtual tool is proposed for the analysis of the operating regimes of an electric power system. Both the block diagrams in which the virtual instruments are built, as well as the front panel with the results corresponding to the root mean square (RMS) values and the voltages and currents waveforms, respectively active, reactive, and apparent powers, are presented here.

Another advantage of using LabVIEW is the fact that it allows the creation of intuitive interfaces in process analysis [18, 19]. Also, LabVIEW can be integrated with MATLAB in order to model, simulate, and test processes [18, 20, 21].

The chapter proposes a LabVIEW virtual instrument that, by means of National Instruments DAQ devices, monitors the operating conditions of a three-phase load and commands a three-phase compensator for power factor improvement.

Within this chapter is presented a short theory necessary to understand the basic theoretical aspects regarding the power factor, respectively, the capacitive compensation of reactive power. Then, we proceed to the presentation of the automatic compensation algorithm and the implementation solution in LabVIEW (architecture of the virtual instrument). In order to test the virtual instrument, a laboratory model was built consisting of a RL consumer and a capacitive compensator in delta connection. The inductive component of the load is adjustable in steps on each phase and is commanded within the automated process. The capacitive compensator is also adjustable in steps, with the possibility of single-phase control. National Instruments equipment compatible with LabVIEW software was used for controlling the adjustable components of the laboratory model and analyzing the process. A stand-alone real-time application was build and deployed on the cRIO 9040 controller in order to realize an embedded operating system.

2. Power factor: Theoretical aspects

Considering a dipolar circuit (**Figure 1a**), the product between the instantaneous values of the voltage at the source terminals, u_s, and the current is called the instantaneous power (Eq. (1)) and can be visualized in **Figure 1b**.

$$p = u_s \bullet i \tag{1}$$

In relation to the active power, it is important to reveal the dependence of this quantity on the phase shift φ. At the same RMS values of voltage and current, the active power varies within wide limits with φ. For example, in **Figure 2a**, the case $\varphi = 0$ was considered, and in **Figure 2b** $\varphi = \pi/2$ was considered (extreme cases). In the first case, $\varphi = 0$ (resistive circuit), the active power is the highest ($P = U_s \cdot I$). It can also be noted that the instantaneous power has only positive values. If $\varphi = \pi/2$ (ideal coil—inductive circuit), the active power is zero, $P = 0$ and therefore, the instantaneous power oscillates between the circuit and the source. The same result is obtained in the case of an ideal capacitor ($\varphi = -\pi/2$).

The RMS value of the active power in a single-phase electric circuit is calculated in Eq. (2).

$$P = U_s \bullet I \bullet \cos\varphi \tag{2}$$

The factor $cos(\varphi)$ that intervenes in the expression of the active power (Eq. (2)) is called the power factor. In Eq. (3) the active current (active component of the

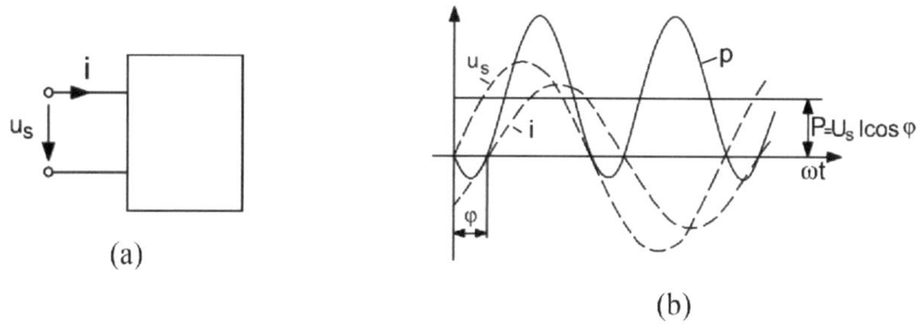

Figure 1.
Dipolar circuit (a) and instantaneous power in an electric circuit (b).

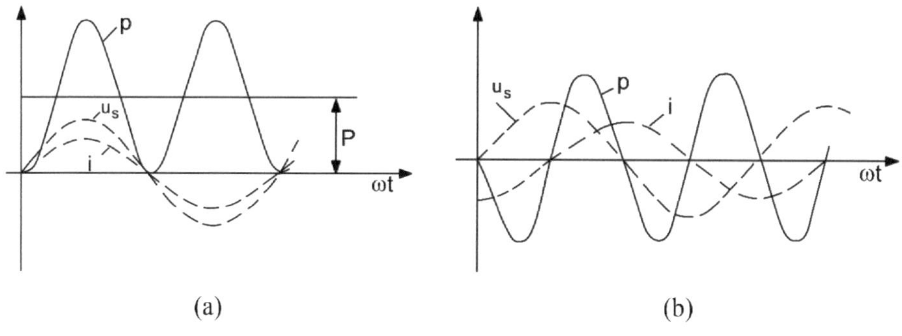

Figure 2.
Instantaneous power in a purely resistive circuit (a), respectively, in a purely reactive inductive circuit (b).

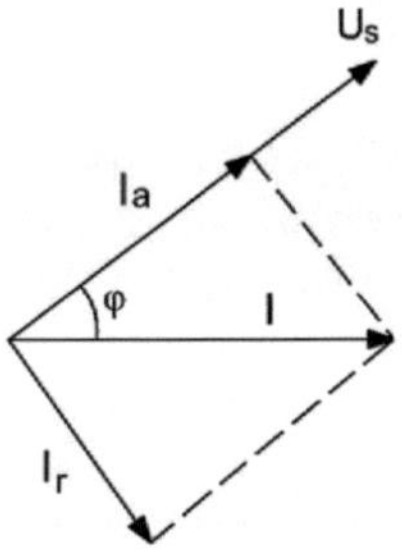

Figure 3.
The phasor diagram of the current.

current) is defined, which in a phasor diagram represents the projection of the current according to the direction of the voltage (**Figure 3**).

$$I_a = I \bullet \cos\varphi \tag{3}$$

The active power can therefore also be written in the form:

$$\mathrm{P} = U_s \bullet I_a \tag{4}$$

The component of the current in a direction perpendicular to the voltage at the terminals (**Figure 3**) is called the reactive component of the current and is defined using Eq. (5).

$$I_r = I \bullet \sin\varphi \tag{5}$$

Based on the reactive component of the current, the reactive power can be defined as follows:

$$Q = U_s \bullet I \bullet \sin\varphi = U_s \bullet I_r \tag{6}$$

In power systems, the power factor should be as high as possible. From Eq. (2), it follows that at a certain voltage at the terminals (the usual case of power supply networks), the same active power can be obtained at different currents. Of course, the solution to obtain a certain power (the required one) with as small current as possible, which means a high power factor, is the rational solution, as it involves small energy losses. A higher power factor can be obtained if the absolute value of the reactive power considered is small.

Unlike active power, reactive power can be positive or negative. Taking into account the rule of signs from receiver circuits, if the circuit is inductive, the reactive power is considered positive and is called inductive reactive power (Q_{ind}). In the situation where the circuit is capacitive, the reactive power is considered negative and is called capacitive reactive power (Q_{cap}).

Increasing the power factor can be achieved through a procedure called reactive power compensation. If an electric circuit absorbs reactive power, it can be compensated (reduced) by supplying reactive power of the opposite sign:

- an inductive reactive power can be reduced by adding a capacitive reactive power; therefore, an inductive electric circuit is compensated by a capacitor bank;

- a capacitive reactive power can be reduced by the contribution of an inductive reactive power, so that a capacitive electric circuit is compensated by coils.

The total reactive power (Q_t) remaining in the circuit after applying the compensation can be defined using Eq. (7)

$$Q_t = Q_{ind} - Q_{cap} \tag{7}$$

In three-phase circuits, the reactive power is calculated as the sum of the reactive powers on the three phases of the circuit:

$$Q = U_1 \bullet I_1 \bullet \sin\varphi_1 + U_2 \bullet I_2 \bullet \sin\varphi_2 + U_3 \bullet I_3 \bullet \sin\varphi_3 \tag{8}$$

The three-phase capacitor bank is composed of an assembly of single-phase units connected together to form a three-phase connection system. From the point of view of connecting single-phase units in a three-phase bank, it is possible to connect them in delta or star. The delta connection scheme has the advantage that, for the same number of capacitors, the reactive power supplied (Q_{CDelta}) is three times higher than in the case of star connection (Q_{CY}). The capacitive reactive power of the capacitor bank is determined according to their connection and thus results the value of the capacitance (C_Y or C_{Delta}) required to be installed in the capacitor bank to achieve total compensation:

$$C_Y = -\frac{Q_{CY}}{\omega \bullet U^2} \tag{9}$$

$$C_{Delta} = -\frac{Q_{CDelta}}{3 \bullet \omega \bullet U^2} \tag{10}$$

3. Hardware device presentation

The virtual instruments developed in LabVIEW allow the use of data acquisition equipment to create real-time applications both in the field of education and in the industrial field. This subchapter presents the hardware equipment used to test the developed virtual instruments.

3.1 NI cRIO-9040 controller

The cRIO-9040 is a rugged, high-performance, customizable embedded controller that offers a dual-core processing. The controller provides precise, synchronized timing and deterministic communications, which is ideal for highly distributed measurements. This controller offers several connectivity ports, which allow to add a local human machine interface and program, deploy, and debug software, which simplifies application development. The chassis of the cRIO-9040 can be used with a combination of C-series I/O modules to create a combination of analog and digital I/O measurements (**Figure 4**) [22].

Figure 4.
NI cRIO-9040 controller.

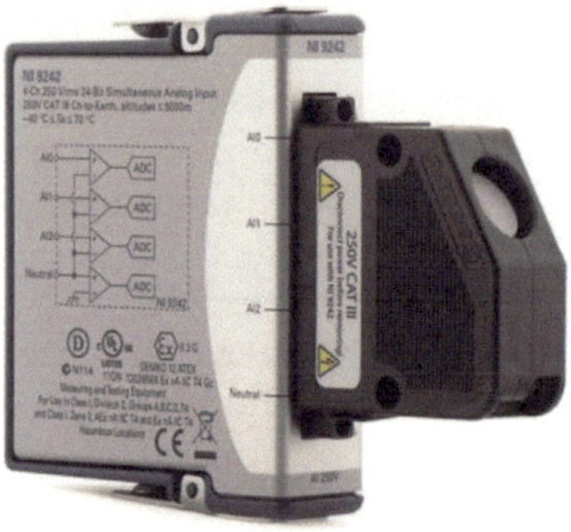

Figure 5.
NI 9242 module.

3.2 NI 9242 module

The NI-9242 module, shown in **Figure 5**, has four analog measurement channels for voltages up to 250 V rms, so it can be used for measurements in single-phase or three-phase systems. The wide measurement range makes it ideal for voltage measurement applications such as phase voltage measurement, power measurement, power quality monitoring. Transient and harmonic analysis can also be performed with simultaneous high-speed sampling [23].

3.3 NI 9227 module

The NI-9227, shown in **Figure 6**, was designed to provide high-precision measurements to meet the demands of data acquisition and control applications. It includes built-in anti-aliasing filters. With four channels of analog inputs for currents up to 5A, if used in conjunction with a C-series voltage input module, NI-9227, it can measure the power consumption and energy for various applications. Different power quality indicators can also be determined: noise, frequency, and harmonics [24].

3.4 NI 9478 module

The NI 9478, shown in **Figure 7**, is a digital output module for CompactDAQ and CompactRIO systems. It features 16 output channels, each channel is compatible with

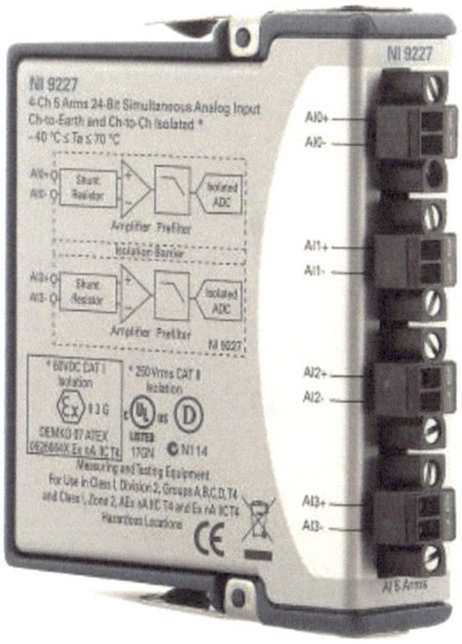

Figure 6.
NI 9227 module.

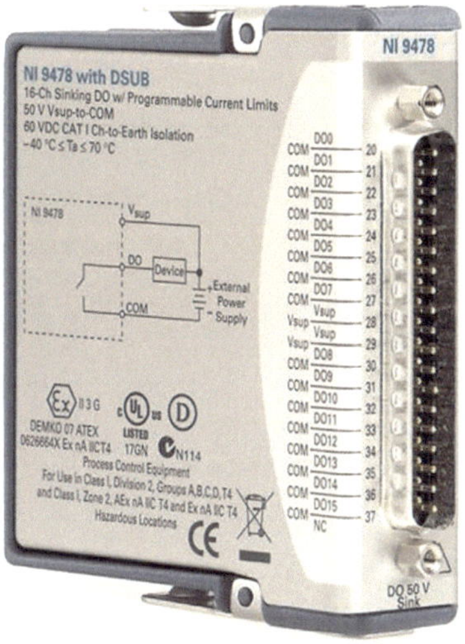

Figure 7.
NI 9478 module.

signals from 0 to 50 V. The NI 9478 works with industrial logic levels and signals to connect directly to a wide range of industrial relays [25].

The output pin numbering of the NI 9478 module is shown in **Figure 8**. Connection to the relays and power supply is made *via* the NI 9923 accessory. This terminal block is designed with a front connection with 37-pin input terminals (**Figure 8**). The NI-9923 facilitates optimal cable positioning and is supplied with a set of jack screws to ensure firm connectivity if intended to be used as part of a high vibration system. The pin terminals of the NI-9923 can be easily accessed after removing the four securing screws of the cover.

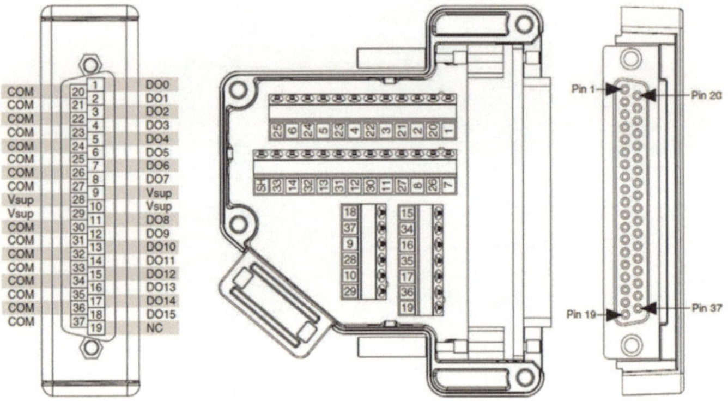

Figure 8.
Output pins of the NI 9478 module and the NI 9923 terminal block.

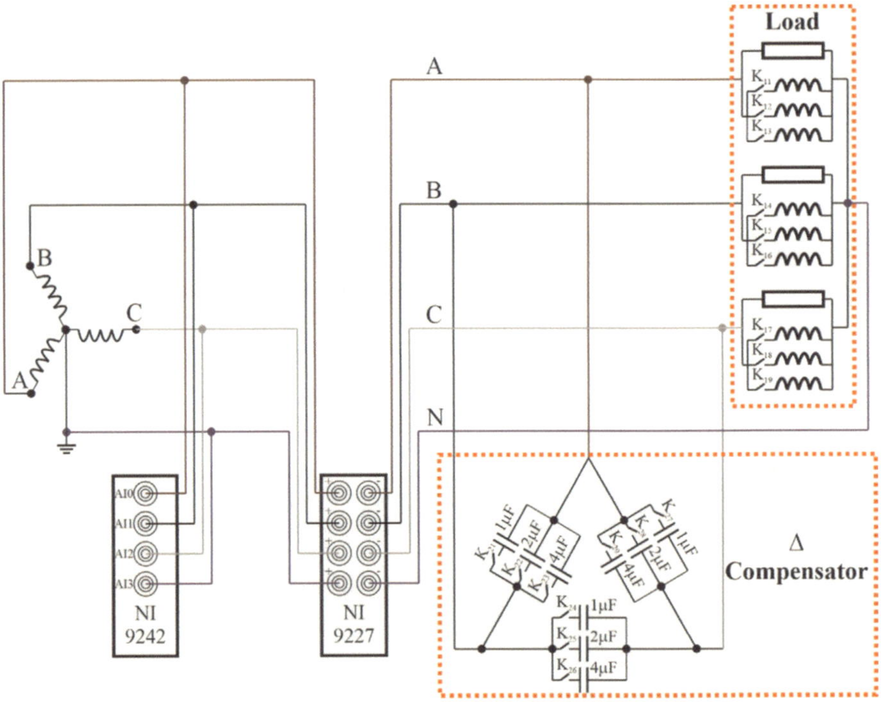

Figure 9.
Modeling diagram of the consumer and the reactive power compensator.

4. The laboratory model built for the compensator control

A relevant example of using LabVIEW in industry and education is the automatic control of a shunt reactive power capacitive compensator. The compensator proposed in this chapter is composed of three compensation steps, in Delta connection, with different values of the capacities: 1 μF, 2 μF, and respectively 4 μF. To test the virtual instrument for compensator control, a laboratory model consisting of an RL consumer in Y connection, powered at 400 V, phase to phase voltage, shown in **Figure 9**, was made.

For both inductive component of the consumer and capacitances of the compensator, the command relays are presented. The correctness of the operation of the compensation algorithm implemented in LabVIEW is verified by monitoring the consumer-compensator assembly, at the common point of connection to the network, using the current and voltage acquired by NI 9227 and NI 9242.

The possible combinations of symmetrical connection of the three steps lead to seven compensation regimes, according to **Table 1**. The table also shows the values of the compensation reactive power values and the equivalent capacity values corresponding to each regime.

The reactive inductive component of the load is step-adjustable, 293 mH, 488 mH, respectively 1465 mH, so as to obtain different values of the power factor and implicitly different values of the compensation requirement. The control of the different steps of the inductive consumer is done from the NI 9478 equipment, according to **Figure 10**.

After the acquisition of currents and voltages by means of the NI 9227 and NI 9242 devices, the power factor and the reactive power required for compensation are determined in LabVIEW. Depending on the result obtained, the NI 9478 device will control the step (or steps) corresponding to the capacitor bank (**Figure 11**).

Regime	The connection variant of the three steps	Q_k [VAR]	C [μF]
1	1	150,796	1
2	2	301,593	2
3	3	603,185	4
4	1 + 2	452,389	3
5	1 + 3	753,982	5
6	2 + 3	904,778	6
7	1 + 2 + 3	105,557	7

Table 1.
Possible operating regimes of the reactive power compensator.

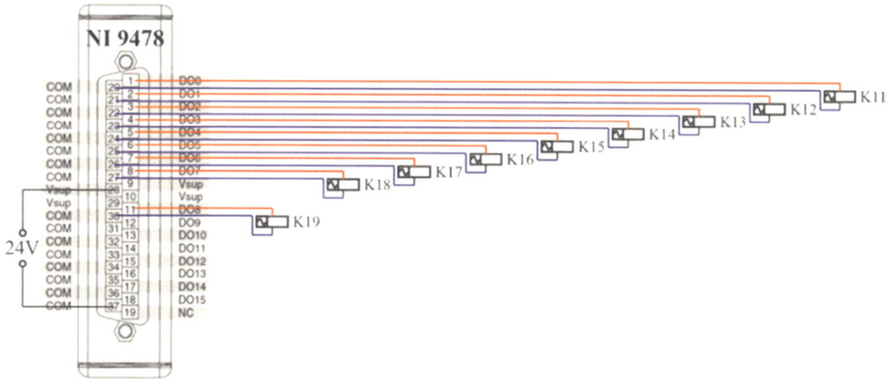

Figure 10.
The control-command diagram of the consumer.

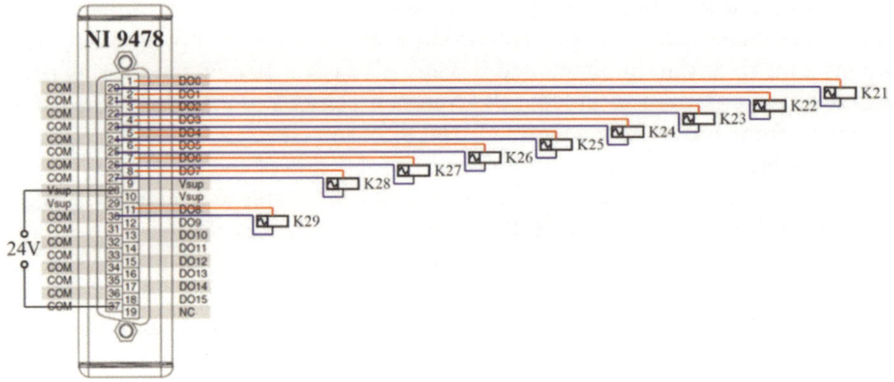

Figure 11.
The control-command diagram of the compensator.

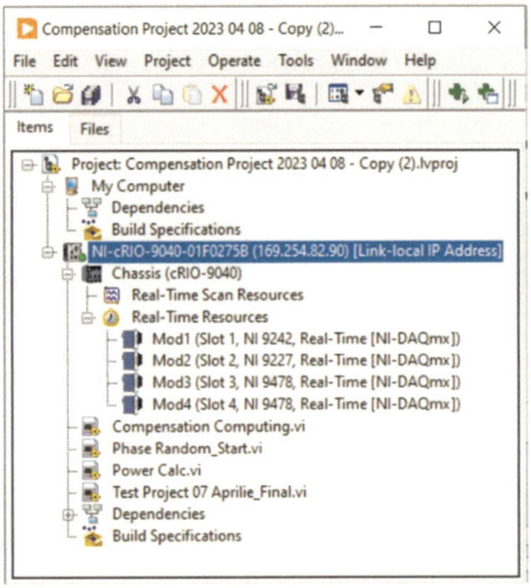

Figure 12.
The compensation project window.

5. Virtual instrument architecture

In order to implement the compensation algorithm, a LabView project was realized and a Real-Time application was build and deployed on the cRIO 9040. In **Figure 12**, the project window containing the real-time resources used is presented.

A *State Machine* design is used to implement the automated process in LabVIEW. The state diagram of the algorithm is presented in **Figure 13**. Here, it can be seen that the algorithm is carried out through five distinct states, which can be labeled *Step 0, Initialization, Process 1, Process 2,* and *Process 3*. In this subchapter is presented the block diagram of the virtual instruments developed with the aim of determining the compensation requirement and the control of the corresponding steps of the shunt capacitive compensator.

The LabVIEW *State Machine* used to apply the algorithm described in **Figure 13** consists of five Case Structures within a While loop, where each of the Case Structure's contains the code associated with one of the state machine's states.

Since it is an automatic process that should execute continuously depending on the load curve, the While loop executes continuously, and the Case structures repeat successively as shown below. The time delay imposed for each case structure is set to in order to be able to follow the development of the entire process and verify the correctness of the algorithm's operation.

a. The purpose of the "*Step 0*" case structure is to switch-off the capacitors each time before the entire process is restarted (**Figure 14**). This step is also useful due to the fact that the application also has a didactic character, and the operation of the algorithm and the development of the process can be easily followed by students.

b. The purpose of the "*Initialization*" case structure is to give a random command for the three steps of the coils corresponding to the three-phase consumer presented in **Figure 9**. *Phase Random_Start subVI* is called here for this purpose (**Figure 15**).

Its main element is the predefined function Random Number (0-1), which returns a random number between 0 and 1. By means of two simple mathematical tricks of

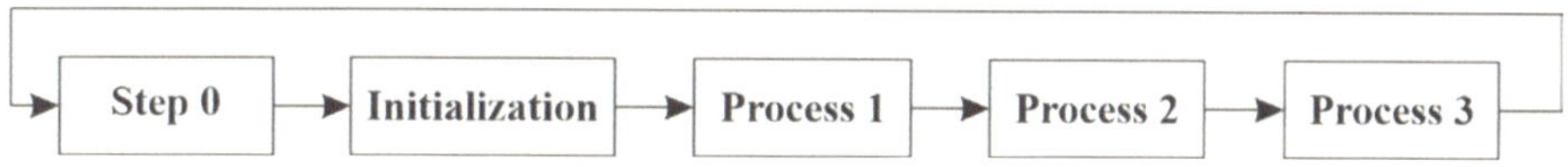

Figure 13.
The state diagram of the algorithm.

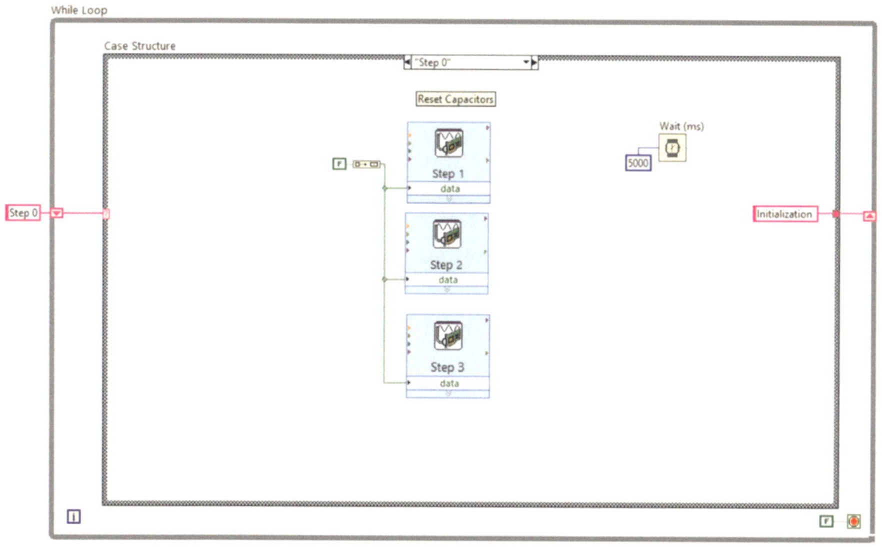

Figure 14.
"Step 0" *case structure Block diagram.*

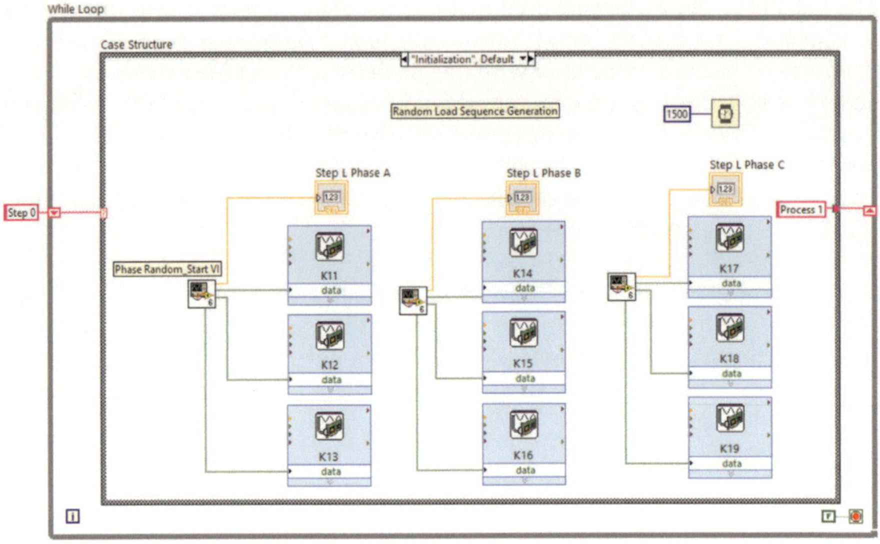

Figure 15.
"Initialization" *case structure Block diagram.*

multiplication and rounding, a random value between 1 and 3 corresponding to the three steps of the coil is then obtained. The procedure is repeated three times, corresponding to the three phases of the network, thus resulting in a balanced or unbalanced consumer depending on the random values provided at the input. When executing the "*Initialization*" structure, among the relays K11...K19, corresponding to the contacts of the three steps of the coils (three on each phase), **Figure 9**, only those corresponding to the generated random values, one for each phase, will receive a switching command *via* DAQ Assistant. The selection is done using the Boolean values resulted by comparison of the generated random values with the constants 1, 2, 3 through the Equal function from *Functions/Comparison*.

c. The "*Process 1*" case structure, **Figure 16**, calls the *Power Calc subVI* and fulfills, first of all, the function of acquiring the voltage and current waveforms on the three phases. It displays the active and reactive powers per phase. The reactive powers are the inductive reactive powers of the load phases. Within *Power Calc subVI,* by means of the predefined Extract Single Tone Information VI, the amplitude and phase of the fundamental frequency current and voltage signals are determined, and then, by dividing the amplitude by $\sqrt{2}$, their rms values. The voltage and current phase values are entered as input quantities in a FOR loop within which, using a Formula Node structure, a procedure for determining the phase shift between current and voltage is applied. This procedure is necessary because the acquired phase values have values in the range [−360, +360] degrees. Having the RMS values of voltage and current and the phase shift between them, the active power, the reactive power, and the power factor can be determined. The procedure is repeated three times, corresponding to the three phases of the network. Taking into account the fact that we propose to compensate the reactive power, also in this case structure the sum of the reactive powers on the three phases is made. The

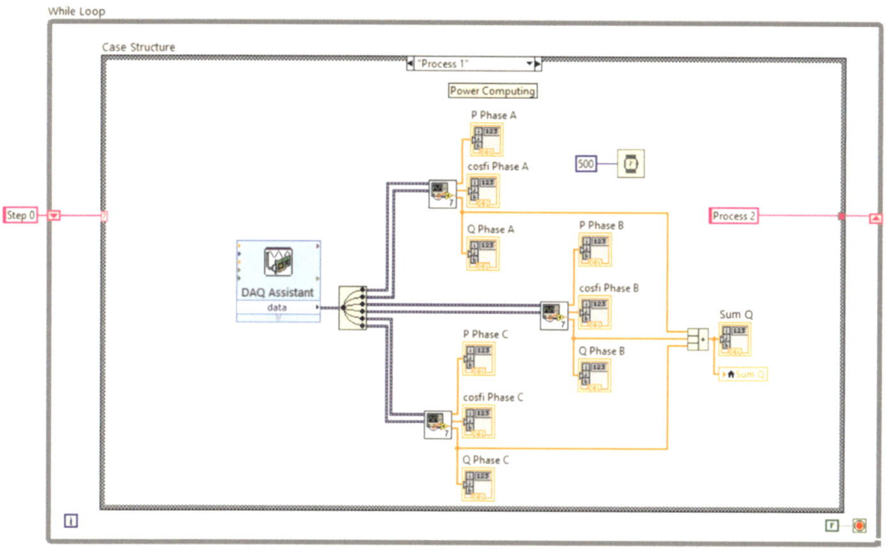

Figure 16.
"Process 1" *case structure Block diagram.*

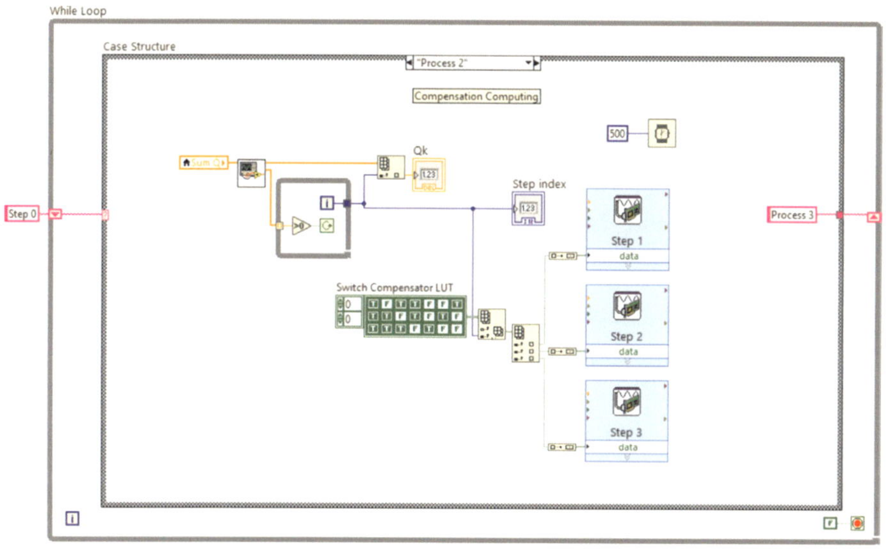

Figure 17.
"Process 2" *case structure Block diagram.*

obtained value will be transferred, by means of the *SumQ* local variable, to the next case structure.

d. Within the Case Structure "*Process 2*," **Figure 17**, the *Compensation Computing subVI* is called and the value of the compensation capacitive reactive power, *Qk*, is determined. The value of the reactive power required for the compensation is read from the local variable *SumQ*, obtained within "*Process 1*" case structure, is compared with the values of the capacitive reactive powers installed on each step of the compensator. The step having the value closest to the required compensation power value, without going into capacitive overcompensation regime, will be connected. For the presented application, there are three steps of capacitors and the possible combinations result in seven steps of compensation. Choosing the right compensation step is done using the *Lookup Table* function in LabVIEW. In this case, the table contains Boolean values, which becomes input data for closing the compensator relays.

e. Within the case structure "Process 3," **Figure 18**, the reactive load is switched off in order to prepare to restart the automatic process.

6. Results

In order to test the correctness of the virtual instrument operating, three balanced and one unbalanced operating conditions are presented. The small differences between the active and reactive values measured on the load phases in the balanced operating regimes are due to the network voltage asymmetries that supply the laboratory model.

Figure 19 presents the powers and power factor values before and after the compensation corresponding to regime 1, the highest value of the reactive inductive power.

Figure 20 presents the voltages waveforms corresponding to the three phases and the phase shift between the voltage and the current corresponding to phase A before and after the compensation.

Being a highly inductive regime, we observe that the voltages waveforms are slightly distorted and also it can be seen that after the compensation, the current is more distorted than before the compensation. In this regime, the capacitive

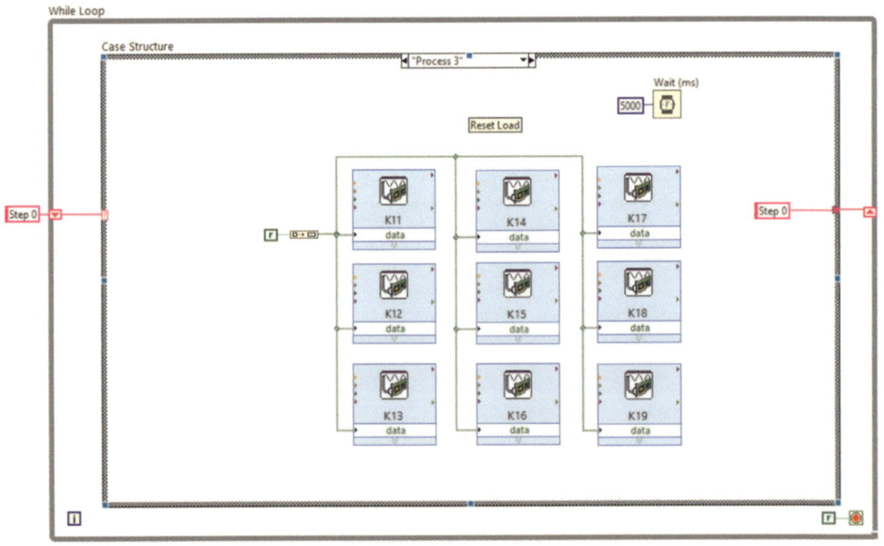

Figure 18.
"Process 3" *case structure Block diagram.*

DOI: http://dx.doi.org/10.5772/intechopen.112669

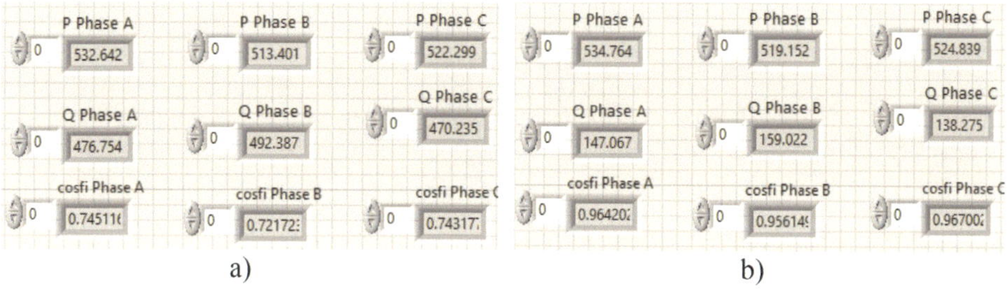

Figure 19.
Regime 1 operating conditions (highest inductive reactive load): (a) before the compensation; (b) after the compensation.

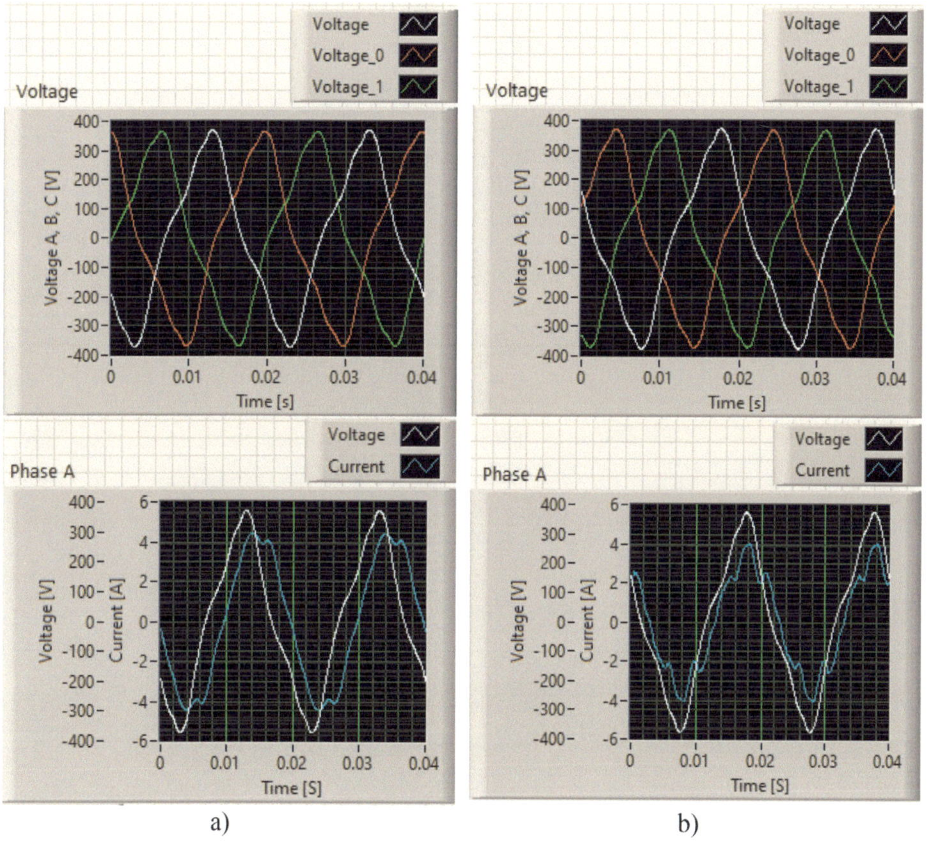

Figure 20.
Current and voltage waveforms—Regime 1: (a) before the compensation; (b) after the compensation.

compensator is operating on the highest step and the power factor after the compensation is about 0.96.

Figure 21 presents the powers and power factor values before and after the compensation corresponding to regime 2, the medium value of the reactive inductive power.

Figure 22 presents the voltages waveforms corresponding to the three phases and the phase shift between the voltage and the current corresponding to phase A before and after the compensation.

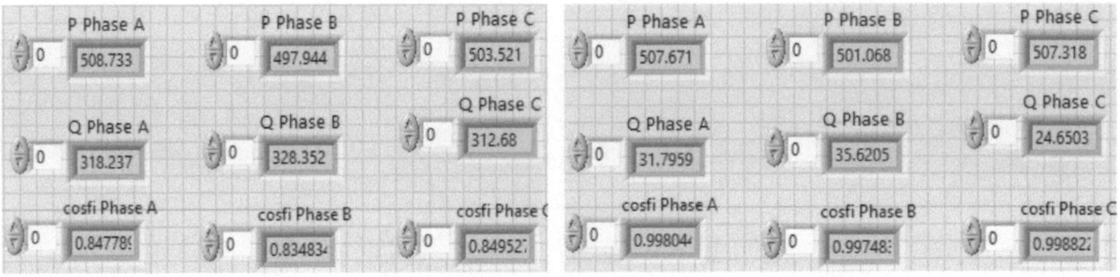

a) b)

Figure 21.
Regime 2 operating conditions (medium inductive reactive load): (a) before the compensation; (b) after the compensation.

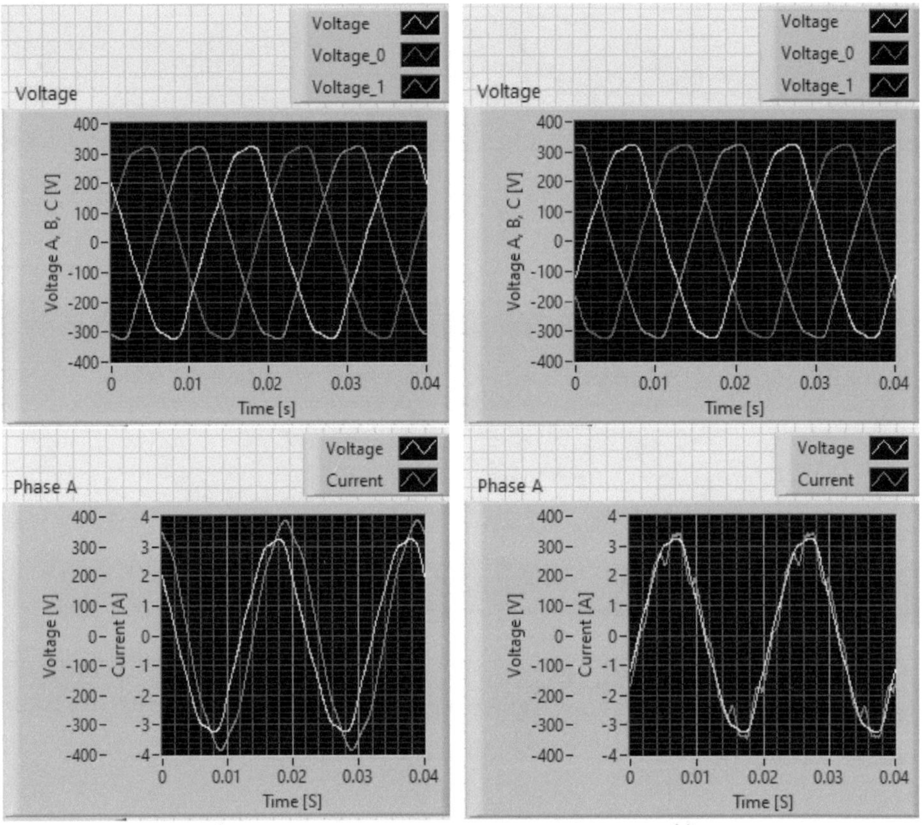

a) b)

Figure 22.
Current and voltage waveforms—Regime 2: (a) before the compensation; (b) after the compensation.

This time, the voltages waveforms are no more so distorted like in the precedent case and the capacitive compensation is total, power factor is practically 1, but the current remains distorted after the compensation due to the presence of the capacitors.

Figure 23 presents the powers and power factor values before and after the compensation corresponding to regime 3, the lowest value of the reactive inductive power.

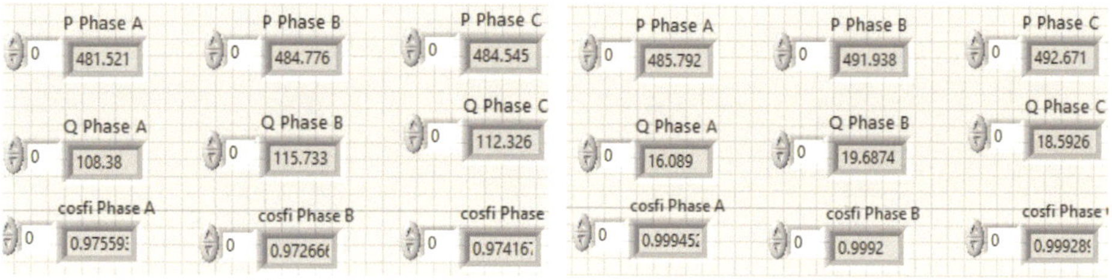

a) b)

Figure 23.
Regime 3 operating conditions (lowest inductive reactive load): a) before the compensation; b) after the compensation.

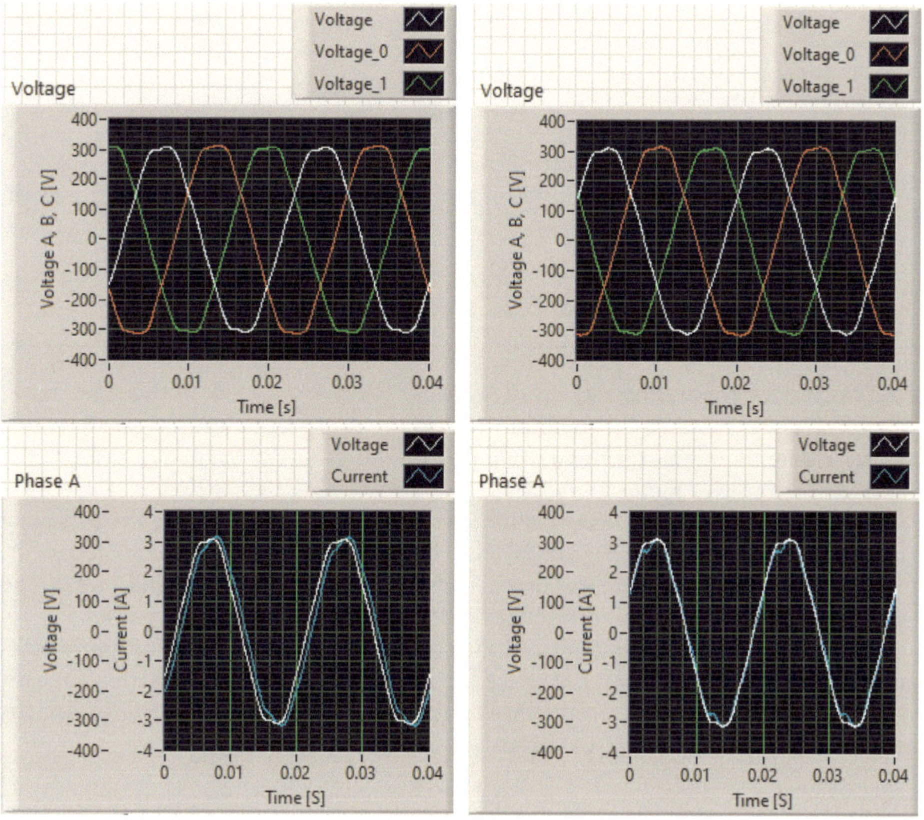

a) b)

Figure 24.
Current and voltage waveforms—Regime 3: (a) before the compensation; (b) after the compensation.

Figure 24 presents the voltages waveforms corresponding to the three phases and the phase shift between the voltage and the current corresponding to phase A before and after the compensation.

In this case, the capacitive compensation is total, power factor is practically 1, and, due to the fact that the capacitive compensator intervenes with the lowest step, the current is almost sinusoidal, without significant distortions.

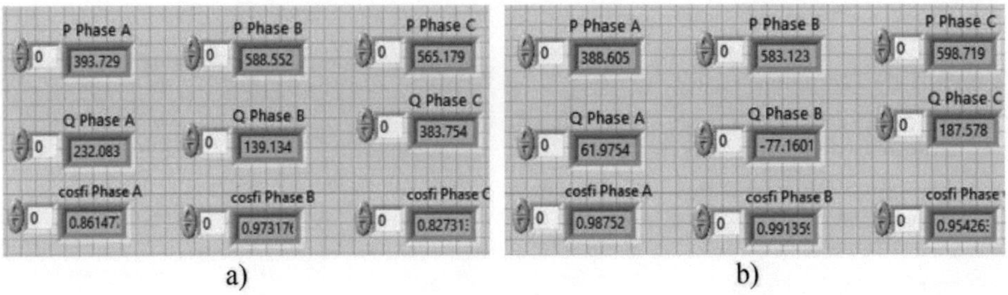

Figure 25.
Regime 4—Unbalanced operating conditions: (a) before the compensation; (b) after the compensation.

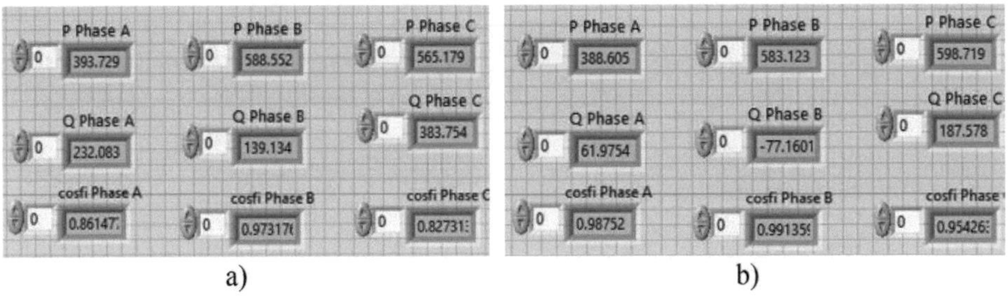

Figure 26.
Current and voltage waveforms—Unbalanced regime, before the compensation.

An unbalanced operating regime (Regime 4) is also presented. In this case, the inductive load operates on the medium step on the A phase, on the lowest step on the C phase and on the highest step on the B phase. The values of the active and reactive powers and the power factor are presented in **Figure 25**.

Figure 26 presents the voltages and the currents waveforms before the compensation. It can be observed in **Figure 26** that the inductive load unbalance causes voltage asymmetries and different power factor on the three phases, and this can be also remarked in the phase shift between voltage and current on each phase.

Figure 27 presents the voltages and the currents waveforms after the compensation. It can be observed that on the phase B is a capacitive overcompensation (negative reactive power), but overall, on the three phases, the total reactive power is inductive (positive), which was the goal of the virtual instrument presented here.

7. Conclusions

Capacitor banks are widespread in electrical distribution networks, this being the cheapest solution for power factor correction.

The accomplishment of a capacitive compensator for power factor correction is not a novelty in power systems, but using LabVIEW software and National Instrument technology offers important perspectives regarding the loads operating regime analysis and optimization. National Instruments technology, through cRIO system and digital output modules, allows obtaining a large number of outputs. This is particularly important when using step switching for large loads that have several steps of capacitors installed.

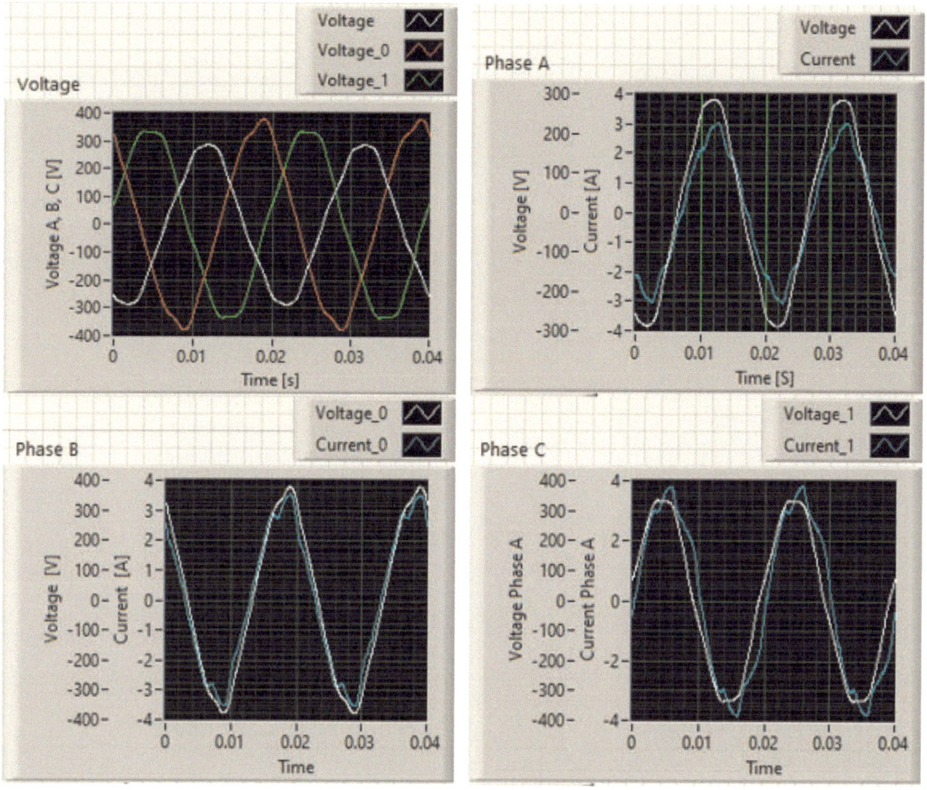

Figure 27.
Current and voltage waveforms—Unbalanced regime, after the compensation.

The solution proposed in this chapter can be implemented especially for unbalanced capacitive compensators designed for load balancing, where the control is done separately on each phase and a larger number of capacitor steps allow a finer control and therefore a better load balancing.

Acknowledgements

This work was supported by a grant of the Ministry of Research, Innovation and Digitization, CCCDI - UEFISCDI, project number PN-III-P2-2.1-PED-2021-4309, within PNCDI III, contract No. 703PED/2022.

Thanks

The authors express their gratitude and thanks to the students from the Master's cycle who, as part of the practical research activity, carried out at the Power Engineering Department within Politehnica University of Timisoara, created the laboratory model for the implementation of the solution presented in the chapter.

Author details

Alexandru Băloi*, Florin Molnar-Matei and Felicia Băloi
University Politehnica Timisoara, Timisoara, Romania

*Address all correspondence to: alexandru.baloi@upt.ro

References

[1] Mahdavi Tabatabaei N, Jafari Aghbolaghi A, Bizon N, Blaabjerg F. Reactive Power Control in AC Power Systems: Fundamentals and Current Issues. 1st ed. Germany: Springer, Power Systems; 2017. p. 634. DOI: 10.1007/978-3-08 319-51118-4

[2] Pana A, Baloi A, Molnar-Matei F. From the balancing reactive compensator to the balancing capacitive compensator. Energies. 2018;**11**(8):1979. DOI: 10.3390/en11081979

[3] Erxia L, Wanxing S, Xiaojun W, Bin W. Combined compensation strategies based on instantaneous reactive power theory for reactive power compensation and load balancing. In: 2011 International Conference on Electrical and Control Engineering. Yichang, China: IEEE; 16-18 September 2011. DOI: 10.1109/ICECENG.2011.6057765

[4] Lee SY, Wu CJ, Chang WN. A compact control algorithm for reactive power compensation and load balancing with static VAr compensator. Electric Power Systems Research. 2001;**58**:63-70. DOI: 10.1016/S0378-7796(01)00127-4

[5] Mayordomo JG, Izzeddine M, Asensi R. Load and voltage balancing in harmonic power flows by means of static VAr compensators. IEEE Transactions on Power Delivery. 2002;**17**:761-769. DOI: 10.1109/TPWRD.2002.1022801

[6] Grünbaum L, Petersson A, Thorvaldsson B. FACTS Improving the Performance of Electrical Grids. ABB Review (Special Report on Power Technologies). Zürich, Switzerland: ABB Group; 2003. pp. 13-18

[7] Dixon J, Morán L, Rodríguez J, Domke R. Reactive power compensation technologies, state of-the-art review. Proceedings of the IEEE. 2005;**93**: 2144-2164. DOI: 10.1109/JPROC.2005.859937

[8] Quintela FR, Arevalo JMG, Redondo RC. Power analysis of static VAr compensators. International Journal of Electrical Power & Energy Systems. 2008;**30**:376-382. DOI: 10.1016/j.ijepes.2007.12.00

[9] Said IK, Pirouti M. Neural network-based load balancing and reactive power control by static VAR compensator. International Journal of Computer and Electrical Engineering. 2009;**1**:25-31. DOI: 10.7763/IJCEE 2009.V1.5

[10] Xu Y, Tolbert LM, Kueck JD, Rizy DT. Voltage and current unbalance compensation using a static VAr compensator. IET Power Electronics. 2010;**3**:977-988. DOI: 10.1049/iet-pel.2008.0094

[11] Haque MH. Compensation of distribution system voltage sag by DVR and DSTATCOM. In: 2001 IEEE Porto Power Tech Proceedings (Cat. No.01EX502). Porto, Portugal: IEEE; 10-13 September 2001. DOI: 10.1109/PTC.2001.964609

[12] Capitanescu F, Wehenkel L. Redispatching active and reactive powers using a limited number of control actions. IEEE Transactions on Power Systems. 2011;**26**:1221-1230. DOI: 10.1109/TPWRS.2010.2102371

[13] Roncero-Sànchez P, Acha E. Design of a control scheme for distribution static synchronous compensators with power-quality improvement capability. Energies. 2014;7(4):2476-2497. DOI: 10.3390/en7042476

[14] Xue Q, Jiang B, Chunxia C, Hang L. Comparison and analysis of reactive

power compensation strategy in power system. In: 2019 IEEE Sustainable Power and Energy Conference (iSPEC). Beijing, China: IEEE; 21-23 November 2019. DOI: 10.1109/iSPEC48194.2019.8975301

[15] Essick J. Hands-on Introduction to LabVIEW for Scientists and Engineers. 4th ed. New York: Oxford University Press; 2019. p. 702

[16] Neeraj K, Sonal J. Development of LabVIEW based electrical parameter monitoring system for single phase supply. In: 2015 Communication, Control and Intelligent Systems (CCIS). Mathura, India: IEEE; 07-08 November 2015. DOI: 10.1109/CCIntelS.2015.7437964

[17] Mohammad A, Ahmad AJ, Tariq EH. Power system analysis using LabVIEW. Universal Journal of Electrical and Electronic Engineering. 2020;7(6): 299-306. DOI: 10.13189/ujeee.2020.070601

[18] Iresha P, Sertac B. Development of LabVIEW based monitoring system for AC microgrid systems. In: 2018 IEEE 12th International Conference on Compatibility, Power Electronics and Power Engineering (CPEPOWERENG 2018). Doha, Qatar: IEEE; 10-12 April 2018. DOI: 10.1109/CPE.2018.8372604

[19] Prabhavathy M, Ramesh B, Kalpalatha Reddy T. An alternative distributed control using LabVIEW. In: 2014 International Conference on Computation of Power, Energy, Information and Communication (ICCPEIC). Chennai, India: IEEE; 16-17 April 2014. DOI: 10.1109/ICCPEIC.2014.6915361

[20] Zhang Z, Jiang H, Zhang H, Wang J. Active power filter design and simulation by combining LabVIEW and Simulink. In: 2011 International Conference on Advanced Power System Automation and Protection. Beijing, China: IEEE; 16-20 October 2011. DOI: 10.1109/APAP.2011.6180971

[21] Lokesh Reddy M, Indragandhi V, Kushal B, Rajasingh R. Integration of MATLAB and LabVIEW for motor control test bench with power analysis. In: 2019 Innovations in Power and Advanced Computing Technologies (i-PACT). Vellore, India: IEEE; 22-23 March 2019. DOI: 10.1109/i-PACT44901.2019.8960064

[22] NI cRIO-9040 Specifications. 2022. Available from: https://www.ni.com/docs/en-US/bundle/crio-9040-specs/page/overview.html [Accessed: June 5, 2023]

[23] NI 9242 Specifications. 2022. Available from: https://www.ni.com/docs/en-US/bundle/ni-9242-specs/resource/376130b_02.pdf [Accessed: September 12, 2022]

[24] NI-9227 Operating Instructions and Specifications. 2022. Available from: https://www.ni.com/docs/en-US/bundle/ni-9227-seri/resource/375101e.pdf [Accessed: September 12, 2022]

[25] NI-9478 Specifications. 2022. Available from: https://www.ni.com/docs/en-US/bundle/ni-9478-specs/page/specifications.html [Accessed: September 12, 2022]

Chapter 4

Automated Measurement of Traction Line Parameters for Railways

Dušan Koniar, Libor Hargaš, Peter Šindler and Marek Paškala

Abstract

In Slovakia, various rail track segments are being rebuilt due to increased railroad speed and safety standards. Assembly wagons were fitted with an automated measurement device to assess the condition and geometry of the traction lines. The theory behind measuring fundamental traction line parameters like line height and stagger is discussed in this book chapter, along with the development of a measurement electronic device with appropriate sensors and a LabVIEW application for evaluating measured data and controlling and interacting with the electronic device. The Railways Research and Development Institute at Vrútky (Slovakia) is where the system was tested and put into use. This work also demonstrates effective and robust collaboration between the academic environment and praxis. The possibilities of using LabVIEW in industrial applications and its cooperation with the developed measuring device will be presented in this book chapter. The possibilities of communication of the device with the developed software and various methods of algorithm design using several structures found in the LabVIEW development environment will be presented.

Keywords: overhead traction line, LabVIEW application, automated measurement, measuring electronic device, railways, data acquisition

1. Introduction

Due to increase in the speed and also the safety of railway transport, many segments of railways are reconstructed and turned into electric in Slovakia. The overhead traction lines and supporting structures are crucial part of electrified railways and must achieve the strict and required geometrical parameters as seen in [1]. The simplified scheme of overhead traction line part is depicted in **Figure 1**.

Actually, there are many systems for measuring the parameters of overhead traction lines. The most often measured geometrical parameters are line height and stagger (horizontal position). Beside electrical parameters, other line parameters can be measured, such as contact wire thickness and vibrations and craters in the contact wire. As mentioned in an older study [2], the systems for measuring the overhead lines can be divided into the following categories: manual or automatic, contact (dynamic) or non-contact (static), and portable or vehicle-borne.

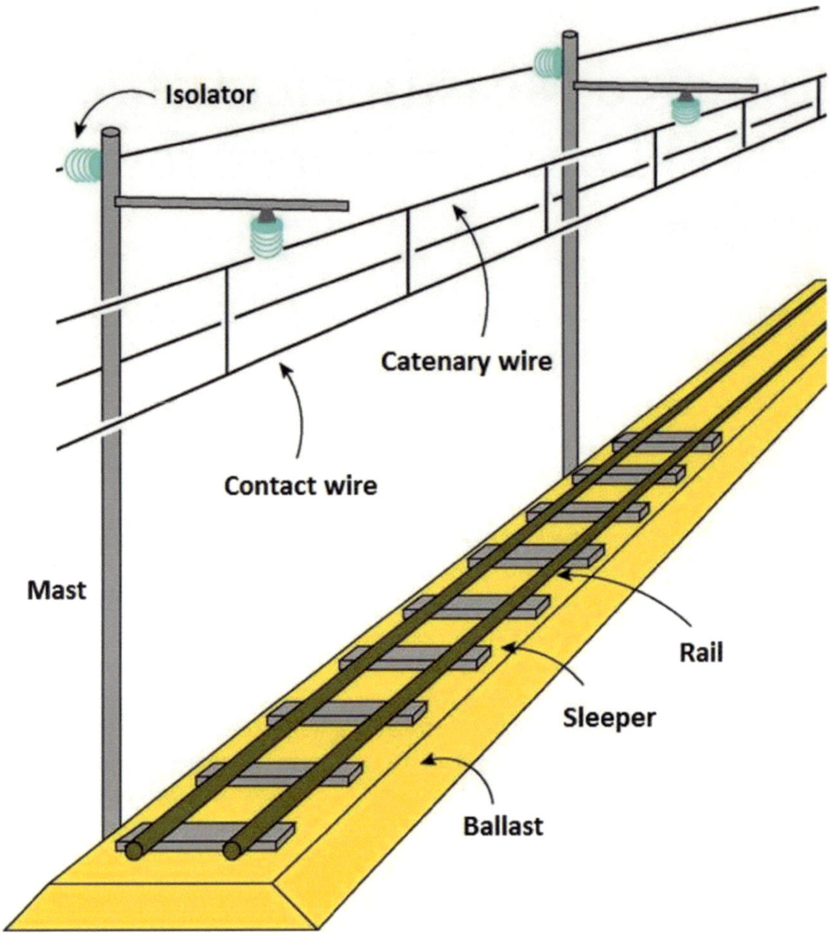

Figure 1.
Structure of railway overhead traction line.

Naturally, non-contact methods are mainly based on usage of optical sensors, instruments, and computer vision algorithms [3–6]. We can also find other measurement approaches using more sophisticated methodologies: laser measurements operating in high-speed trains in Japan [7], LiDAR [8, 9], or radar [10] systems.

Contact methods use a series of sensors built in a locomotive pantograph. The measurements are based on contact forces between a pantograph and a contact wire [11]. The contact methods often measure the condition of a pantograph strip and the quality of interaction between the pantograph strip and the contact line [12, 13]. Many measuring systems are offered as commercial products (e.g., Tesmec or HBK).

For continuous measuring and simplification and objectification of the height and stagger of overhead traction lines, we developed a sensor system built in the pantograph of an assembly wagon linked with the central electronic device. The entire system and measurement procedures may be controlled by the LabVIEW application. The advantage of this system is automatic data acquisition and storage in the system internal memory.

The proposed electronic device has two operating modes: autonomous mode and PC control via specialized service software (LabVIEW application). Evaluation software can be used to analyze all measured data extracted from the internal device memory. When assessing the parameters of overhead traction lines, all the national technical standards are respected. The advantage of LabVIEW-based application design is that the measured data (tables and graphs) are presented in graphical user interfaces with a professional look. If some parameters exceed the defined tolerance, an overrun table is generated. The LabVIEW application also supports the printing of the measurement report.

2. The geometry of the measurement system

To measure the two basic parameters of overhead traction line geometry (contact wire height and stagger), the basic geometric properties and relations must be known. For measuring the heights, wire position sensors are used. For stagger computation, the photodiode array is implemented parallel to the pantograph strip. The simplified geometry of measurement is depicted in **Figure 2**. As shown in this figure, the resulting contact wire height is composed of several partial heights, and information from position sensors (measuring, e.g., the wagon vertical deviations) is dedicated to compute the corrected (perpendicular) height.

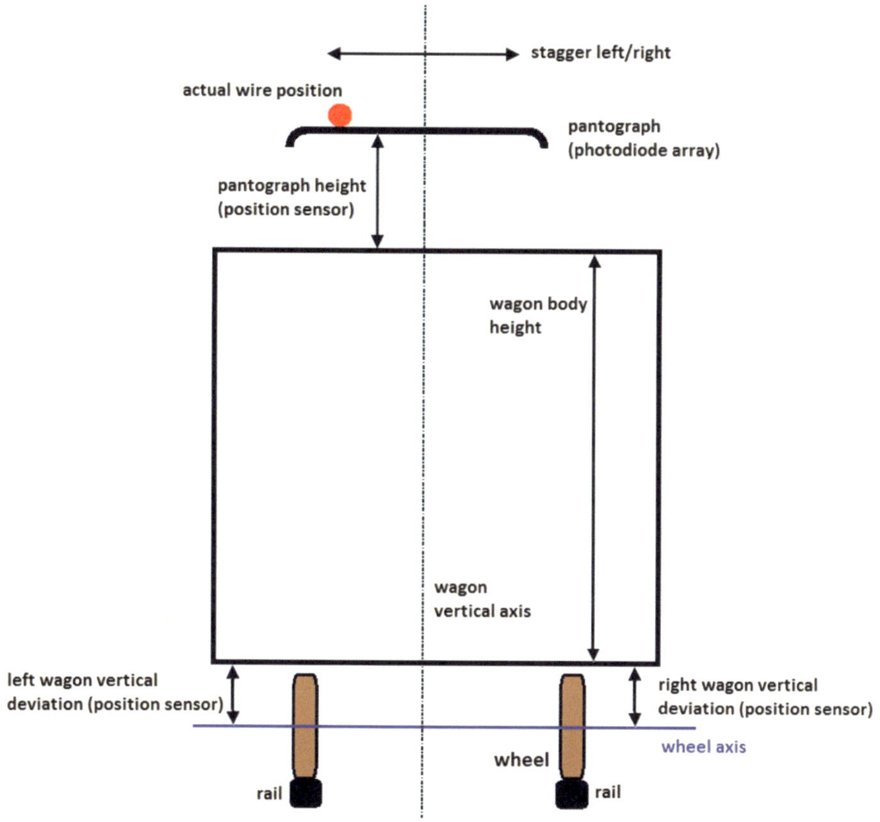

Figure 2.
Geometry of system used for parameter computation (the schematic is not in a proper scale; it serves only for visualization of geometric relations).

Figure 3.
The wire position sensor measuring the vertical deviation of the wagon body from the wheel axis.

The following conditions are introduced in the measurement: the wagon body is hard, and in turn, the wagon vertical deviations are measured with wire position sensors (**Figure 3**). Also, the height of the pantograph (measured from the roof of the wagon) is measured with a wire position sensor.

Deviations are used for computing (correcting) the real height of the wire. The final height is a sum of partial (corrected) heights from the rails to the pantograph strip. Contact force between the pantograph and contact wire is kept as constant and controlled by an independent force sensor. The pantograph sensor for stagger measurement is shown in **Figure 4**.

The accuracy of height measurement is set under 1 cm, while other parameters are under 1 mm.

The last important part of the entire measurement system is the GPS module, which enables fusion of the measured overhead line data with the physical track. This data fusion is also possible due to the implemented trajectory increment sensor. The trajectory increment sensor is placed on the wheel and can measure one-hundredth of wheel turns. The wheel perimeter is approximately 2560 mm. The device can capture each data from this sensor, but the default increment is set to 4. It means that the device allows calculating track line parameters that are 10.24 cm each. This is a sufficient distance to calculate the parameters according to norms. The position of masts, as well as special events, is stored and aligned with the GPS position.

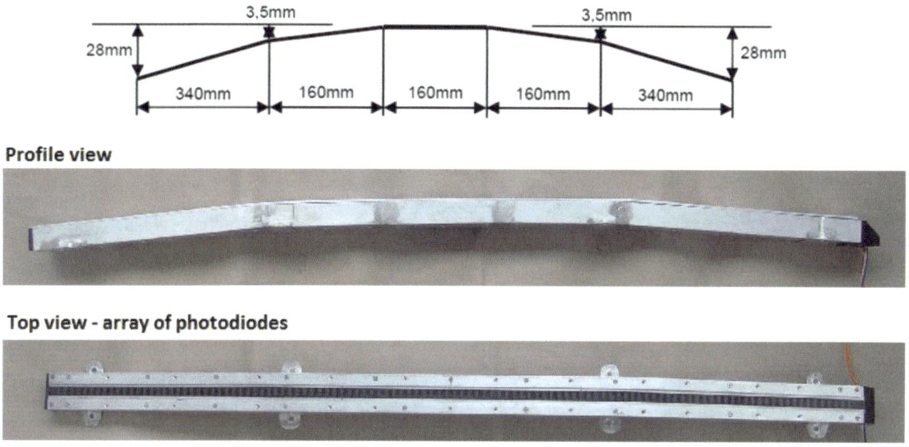

Figure 4.
The pantograph sensor built in an aluminum skeleton: physical dimensions, profile view, top view—the array of photodiodes.

3. Acquisition device

An acquisition device is a complex electronic system developed for setting the measurement sensors, reading the data from sensors, and storing the measured data. The physical device is shown in **Figure 5** and block diagram in **Figure 6**. An acquisition electronic device is located in the cockpit of the wagon.

3.1 Physical properties and parts of an acquisition device

The main physical components of acquisition devices are:

- motherboard/microcontroller,
- detector of contact wire position on the pantograph strip,
- GPS signal detector,
- wagon deviation detector (right—A and left—B),
- trajectory incremental sensor,
- interconnection optical cables.

The 32-bit CPU used in the microcontroller core is from the NXP manufacturer. The microcontroller simultaneously interacts with other detectors and sensors as well as with the position of the overhead traction line on the pantograph strip detector using optical interconnection on a periodic basis. The display and keyboard are also connected via the ports. The PC is connected to the device via a USB port. The measured data can then be sent and analyzed on the PC with the LabVIEW application. Selecting measurement settings, starting, or terminating is possible from this connected PC.

Figure 5.
The physical appearance (front panel) of an acquisition device.

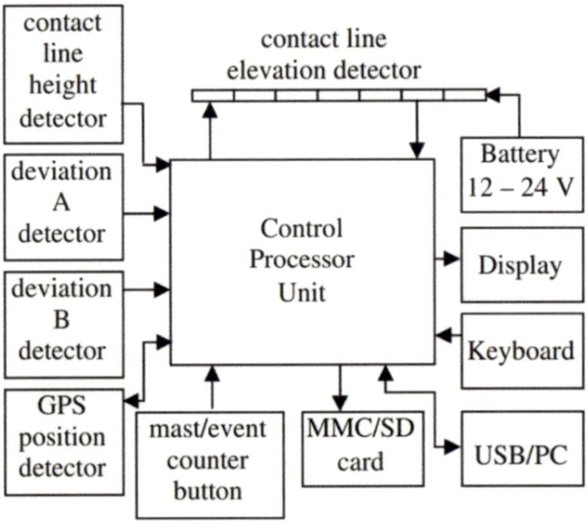

Figure 6.
The block diagram of an acquisition device.

3.2 Software features of an acquisition device

It is possible to immediately evaluate the parameters that were measured. The device and PC can communicate online. The software features of an acquisition device are:

- online numeric and graphic values displaying on the device's display,
- online numeric and graphic values displaying on the PC through the USB cable,

- measurement settings entered from the device keyboard,
- measurement settings entered from the PC through the USB cable,
- device settings entered from the device keyboard,
- the measurement can be started from the device keyboard or from a PC through the USB cable,
- measurement overview on the device,
- measured parameter recording,
- saving the record to the MMC/SD card,
- list of last 20 measurements,
- password-oriented access to the device (tree level),
- the automatic actual time and date setting,
- measured data evaluation on PC.

3.3 Acquisition device motherboard

Dual-layer board is how the motherboard is implemented (1.5 mm). The following building blocks are included: the processor unit, the 16-bit converter with analog multiplex, the power supply component, the binary signal inputs, the MMC/SD card, the USB controller, the serial optical interface for the position of the overhead traction line on the pantograph strip detector, the GPS module, and the RTC controller.

On the motherboard is the 32-bit LPC2214 with the ARM7TDMIS core [14]. The frequency of operation is 55 MHz. For programs and other peripheral controllers, this CPU has 256 kB flash memory in addition to 16 kB RAM. The CY62167DV30 external 16 Mbit integrated RAM is added to the processor. The operating system uses this memory.

For storing measured data, an external NAND memory with a 1 Gbit capacity is used. The entire recording capacity is between 200 and 240 km. This amount depends on the measurement precision you've set. The increment count from the trajectory incremental detector determines the measurement precision.

A common 16-bit bus connects each circuit. The programmable logical field XC2C64A and the processor work together. With the help of this logical field, the following interfaces can be realized: keyboard, binary inputs, serial controller, and PC connections.

3.4 Position sensor of the contact wire

The sensor is designed as a non-contact sensor (photodiode) working on the principle of reflection of infrared radiation from an obstacle (overhead traction line) (**Figure 4**). An integrated photodiode OPB732 is used, which emits infrared light with a wavelength of 850 nm [15]. The sensitivity indicated by the manufacturer is a maximum of 5.5 cm, but 2.54 cm is recommended.

The sensor works in the dynamic (impulse) mode. There are 16 integrated gates on one module at distances of 1 cm. The position sensor consists of seven modules. The number of modules is optional—it is set in the menu of the control unit.

4. The PC's software description

The software solution for measuring overhead traction line parameters is created in the LabVIEW development system. The application was created for a more user-friendly environment for communication with the hardware part (**Figure 7**). With the help of the created software, it is possible to perform three basic modes when measuring parameters:

1. setup

2. online measurement

3. evaluation

For the first two modes, it is necessary that the PC with the software is connected to the hardware part using a USB connection. The third mode is used without the need to connect to the device. It should be noted that the first two modes can also be implemented directly in the device using the keyboard without the need to connect to a PC.

The selection of the application mode is realized as a state machine. When the mode is selected, appropriate tabs must be displayed (**Figure 8**). The user interface contains the "Tab control," and for different modes, different tabs must be displayed. The state is transferred through the local variable to the next parts of the block diagram.

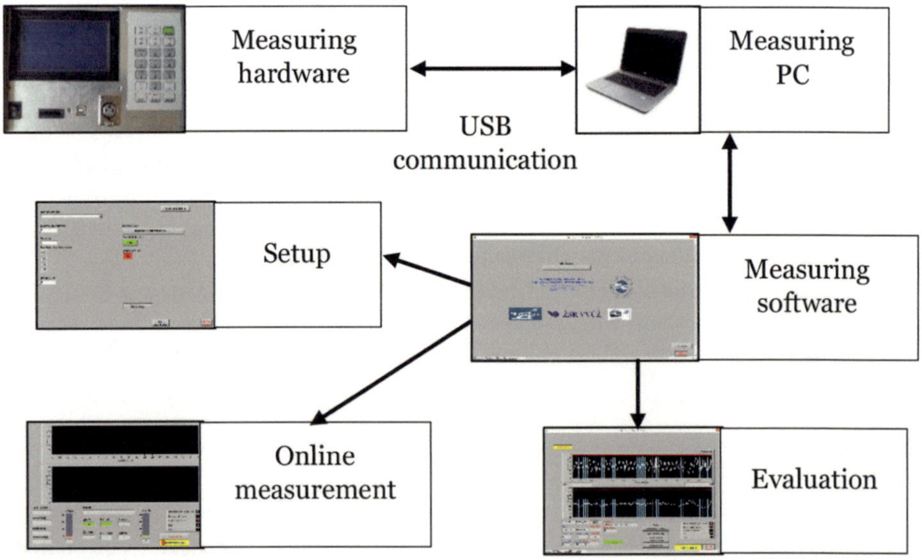

Figure 7.
Hardware (device and PC) and software (setup, measurement, and evaluation) blocks that are needed for the right functionality of the device.

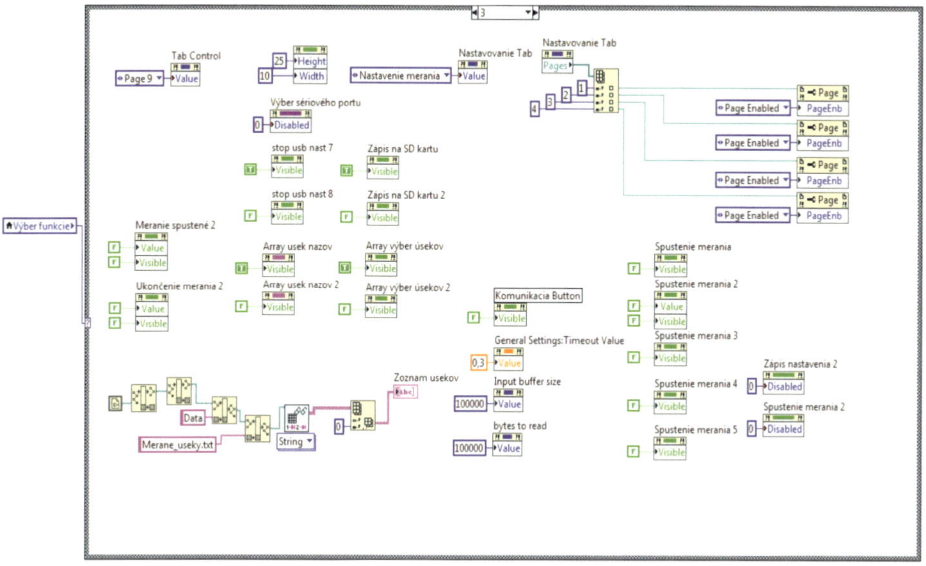

Figure 8.
The block diagram for mode selection.

4.1 Setup

In the case of selecting the "Setup" mode, the PC must be connected to the hardware part. In this mode, the device operator can set the input data for measuring the overhead traction line parameters (**Figure 9**). If the measurement in the given section has not yet been carried out, it is necessary to enter all the necessary data for the measurement of this section. The name of the measured section can be considered the basic data. This name is important as it is subsequently used in the evaluation of data from this section. The name of the measured section is used when creating the names of the evaluated files. It is also used when comparing the results of measuring parameters for multiple measurements of the same section.

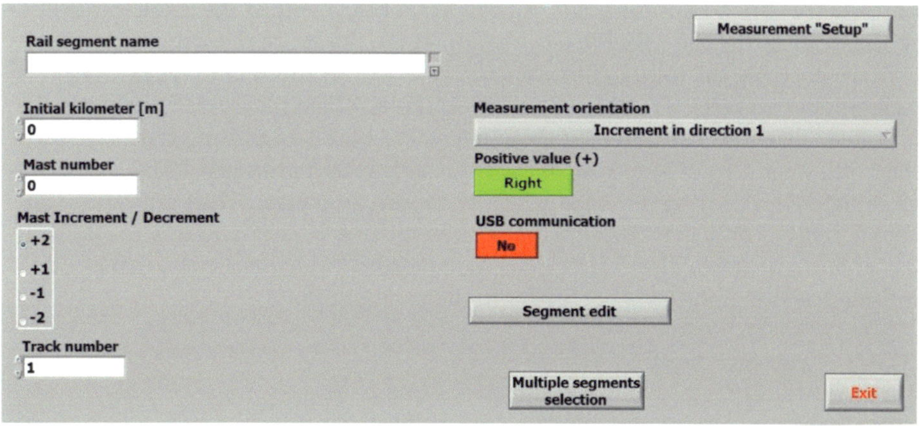

Figure 9.
The front panel of the "Setup" mode.

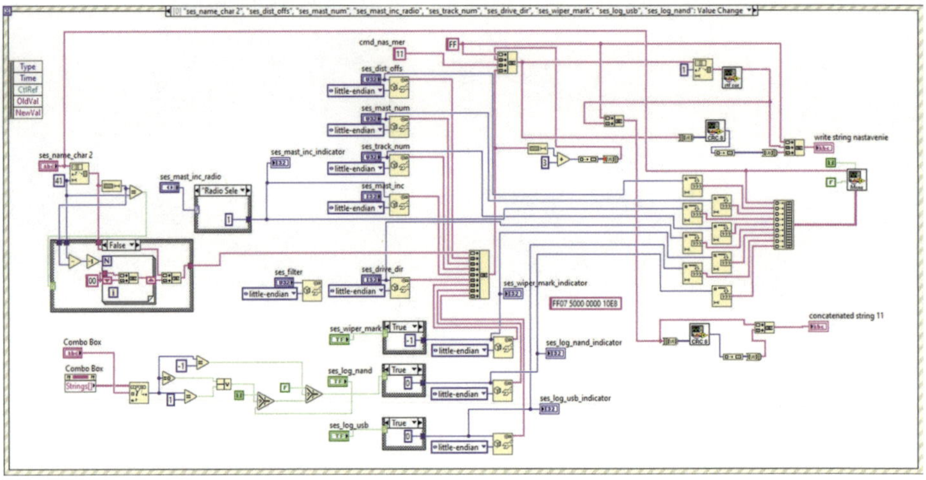

Figure 10.
The parameters are written in the "Event" structure.

Users can type all measurement parameters at once, and these parameters will be written to the variables after the corresponding button is pressed. The "Event" structure in the block diagram is used to provide this functionality (the example is given in **Figure 10**).

Other data that are set in this mode are the initial mast, initial kilometer, and track number. The entered data serve as initial values when the parameters are measured. With the initial mast, it is also necessary to set if the number of mast will be increased or decreased. This data is used for manual marking of masts located on the measured section. When measuring, the location of the masts is marked with a joystick button. Since the information about the initial mast is entered as numerical information, it is possible to increase or decrease the number of the measured mast by the set constant. In many cases, however, in the measured sections, the masts are marked not only with a numerical value but also with a letter. After the first measurement of the section, these masts can be edited during the evaluation (more information is given in Section 4.3). If a list of masts is available for the measured section, the mast designation is read from this file. In this case, it is important that the name of the measured section and the name of the file with the designation of the masts are the same.

When entering the initial kilometer, the value corresponding to the railway line kilometer at the first measured mast is entered. When measuring the section, the crossed distance is then increased or decreased based on the orientation of the measuring wagon and the setting of the trajectory increment sensor. The last important input data is the stagger value at the first measuring mast. The exact value is not entered, but it is only mentioned whether it is positive or negative from the point of view of measurement. The exact value is determined based on the position of the overhead traction line above the sensor in the pantograph.

4.2 Online measurement

The "online measurement" mode is used to display currently measured parameters. To use this mode, it is necessary that the PC is connected to the hardware via a USB. At the same time, data transfer via USB must be enabled in the "Setup" mode.

DOI: http://dx.doi.org/10.5772/intechopen.108873

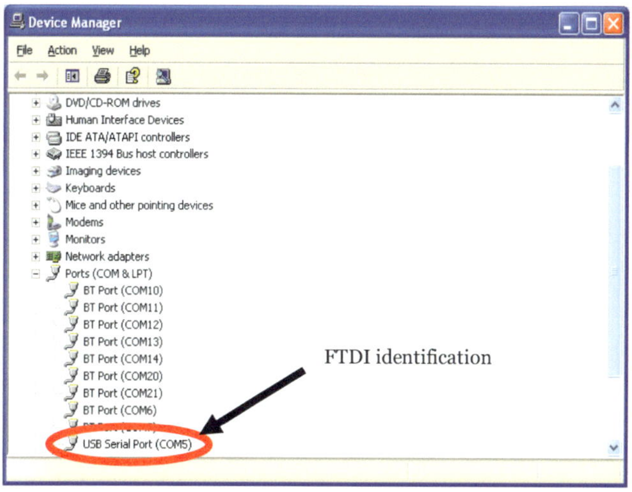

Figure 11.
Viewing the device in the "Device Manager" in the Windows operating system.

The acquisition device implements an FTDI chip FT2XX that can provide serial communication. The acquisition device is presented in Windows "Device manager" as an additional COM port (**Figure 11**). The application checks the connected COM devices and tries to find the device with the FTDI description. The application increments the COM port number in the while loop and searches the string "USB," which is typical for a connected FTDI chip. The subVI for searching for a connected COM port is given in **Figure 12**.

After the device is found, communication can be established between the device and the PC in two ways. The VISA functions can be used, and these functions are implemented in the LabVIEW environment, or the FTDI D2XX function library for LabVIEW can be used, but this library must be downloaded and incorporated into the environment. An example of communication between the PC and the device through implemented FTDI functions is shown in **Figure 13**.

When FTDI functions are used, the state machine is mainly designed for writing and reading data. The individual steps of communication setup are designed as separate states. After starting the communication via FTDI, it is then sufficient to refer to the reading or writing state. The implementation of the state machine therefore makes the block diagram more transparent and simplifies it.

By selecting this mode, after synchronization, the data on the measured parameters will be displayed on the PC screen in both graphic and numerical

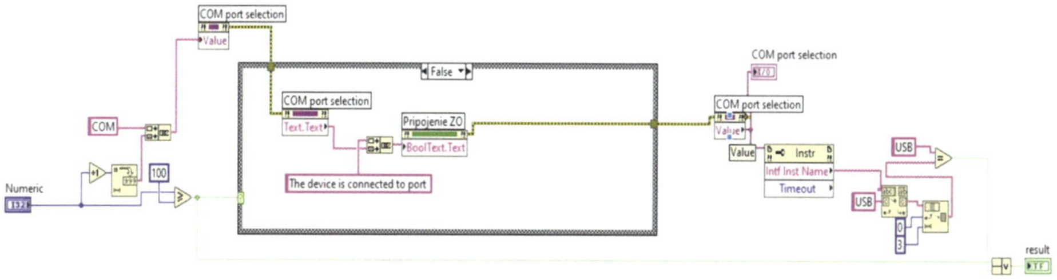

Figure 12.
The code for searching for a connected COM port.

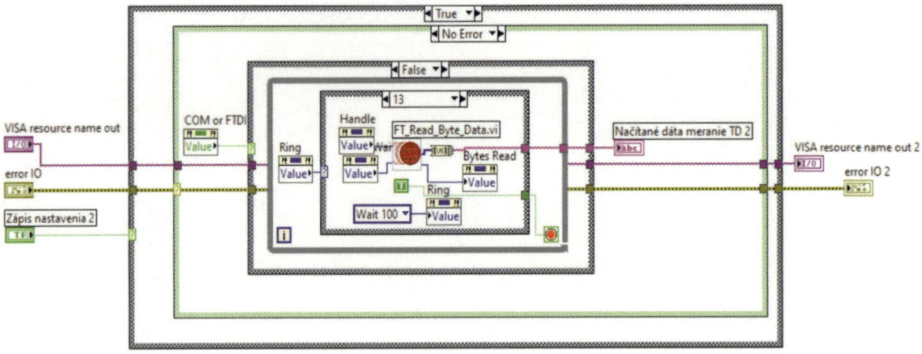

Figure 13.
The state machine with FTDI subVIs is used for the creation of the USB connection.

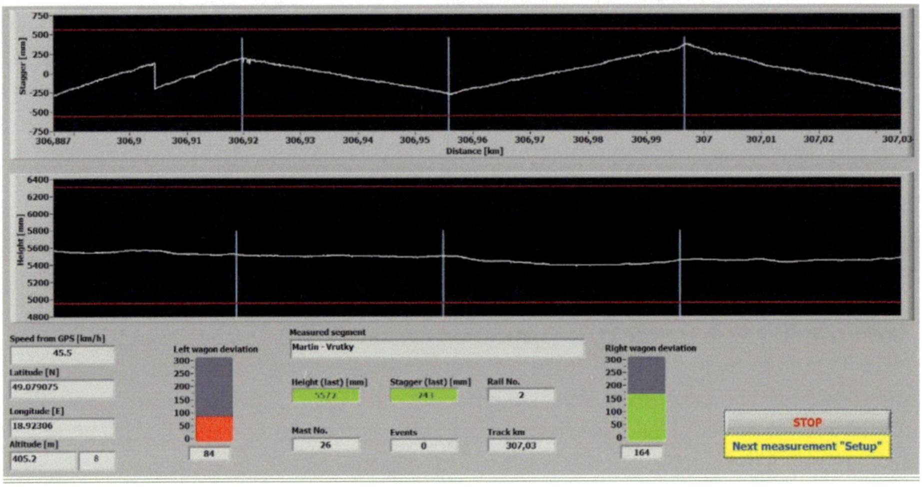

Figure 14.
The front panel of the "Online measurement" mode.

forms (**Figure 14**). Individual indicator elements are used not only to display the status of the measured parameters of contact line height and stagger but also to indicate the status of auxiliary data from the sensors. Auxiliary data includes, for example, information about the deviation of the measuring wagon. This parameter is important in terms of contact line height and stagger correction. Along with data on currently measured values, data on the measured section is also displayed on the screen. There is information about the crossed distance, the current mast field, and the number of the measured railway track.

Only the last 100 m of the measured data for height and stagger measurement are displayed on the graphic indicators. The coloring of the numerical indicators indicates if the limit values have been exceeded (red—exceeded, green—within the limit). The next displayed information is information obtained from the GPS module—actual speed, latitude and longitude, and altitude.

4.3 Evaluation

This mode is used if the measured data about the overhead traction line is saved on the SD/MMC card. During the first processing of the measured data files, the evaluated data are written into table files that can be viewed with table editors. When the data file is subsequently loaded, the preprocessed data stored in the table files are loaded directly from this table file.

The measured data are written in a binary data file. The data from sensors are written directly to the file in the raw form as U32 data type. Therefore, it is necessary to convert these data into real values. Real values in the DBL data type are achieved by incorporating the entered constants and the position of the contact wire. An example of conversion of binary data from sensors to real values is shown in **Figure 15**.

During the first evaluation of the measured section, it is possible to edit some measured data (**Figure 16**). These data need to be adjusted because they will be used as a reference when measuring the same section repeatedly. When editing the data, a new window will appear. In this window, masts of overhead traction line can be edited. The user can insert or delete a mast. This situation occurs if the operator incorrectly marked the location of individual masts during the first measurement. The place of insertion or deletion of the mast is indicated by moving the cursor to the place where the user wants to insert the location of the mast. Mast editing also serves to change the generated mast number. This option is possible if the list of traction masts for the measured section is not known in advance.

When measuring the same section repeatedly, this option will no longer be available. Masts will be automatically added to the evaluated data based on the first accurate measurement.

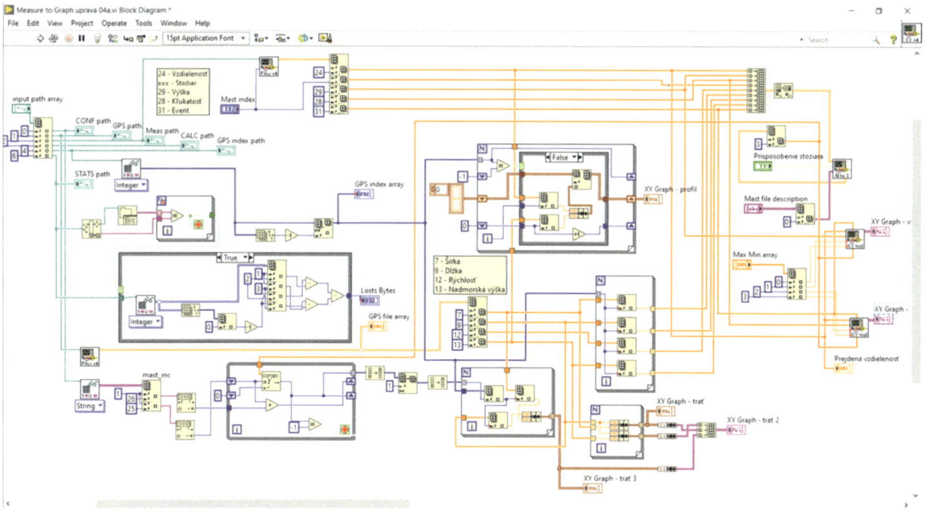

Figure 15.
The block diagram of the "Evaluation" mode where the binary stored raw U32 data are transformed to readable numbers DBL.

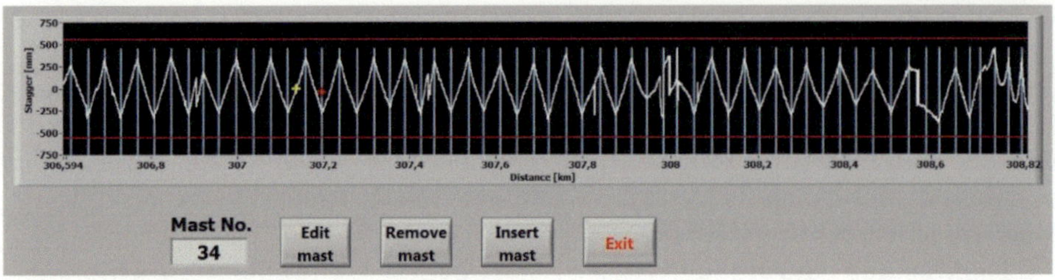

Figure 16.
The front panel of the mast adjustment for initial evaluation.

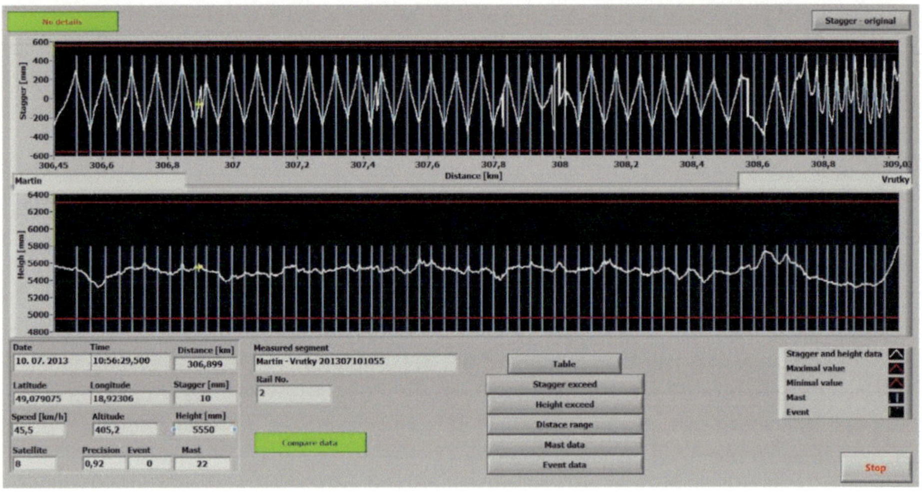

Figure 17.
The front panel of "Evaluation" mode.

Browsing the measured data in the displayed graphs can be done only after evaluating all the measured data (**Figure 17**). After displaying the measured data, it is possible to find out the height and stagger data using the cursors located in the graphs. At the same time, data corresponding to the position of the cursor in the graph is displayed in the numerical indicators. The displayed data are:

- distance—the value is given in km,
- stagger—the value is given in mm,
- height—the value is given in mm,
- mast number,
- GPS data:
    - date of measurement,
    - measurement time,

- latitude,
- longitude,
- actual speed during measurement,
- altitude,
- number of satellites,
- accuracy of measuring the position of the measuring device (wagon).

In this graphic display, it is possible to view the measured data for the entire measured section, or it is also possible to choose a detailed display of the measured data. After selecting the detailed display, it is possible to display individual measured sections either according to the crossed path or as sections between individual masts (**Figure 18**).

When evaluating the data, there may be inaccuracy in determining the parameters. These cases often occur when overhead traction lines are crossed. It happens when one overhead traction line ends and the other starts. Such a crossing will be reflected in the different heights of the overhead traction line. Therefore, it is possible to display the original data obtained from the sensors to accurately determine the parameters (**Figure 19**).

Since the measuring system is primarily intended for determining the parameters of height and stagger, it is necessary to detect the exceeding limit values of these parameters. The software allows the user to enter the values of the limit parameters.

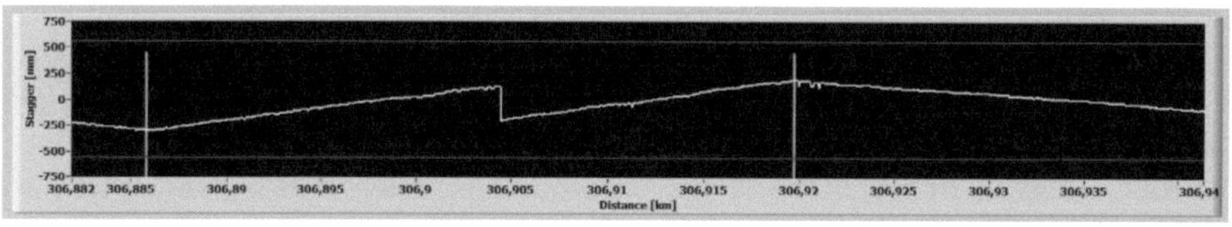

Figure 18.
Detailed display of stagger between two masts.

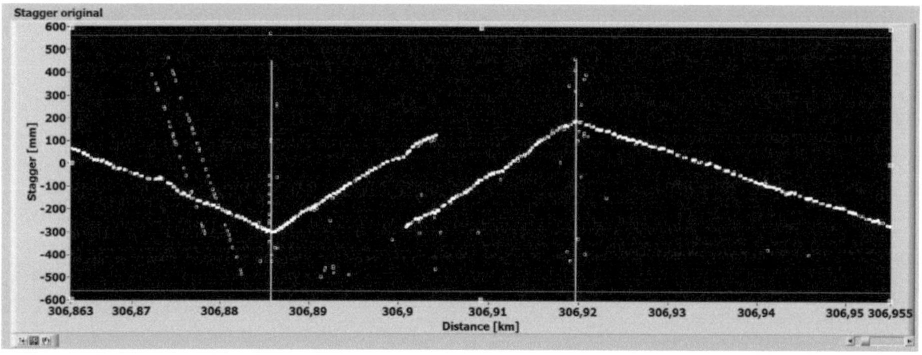

Figure 19.
Original data from sensors.

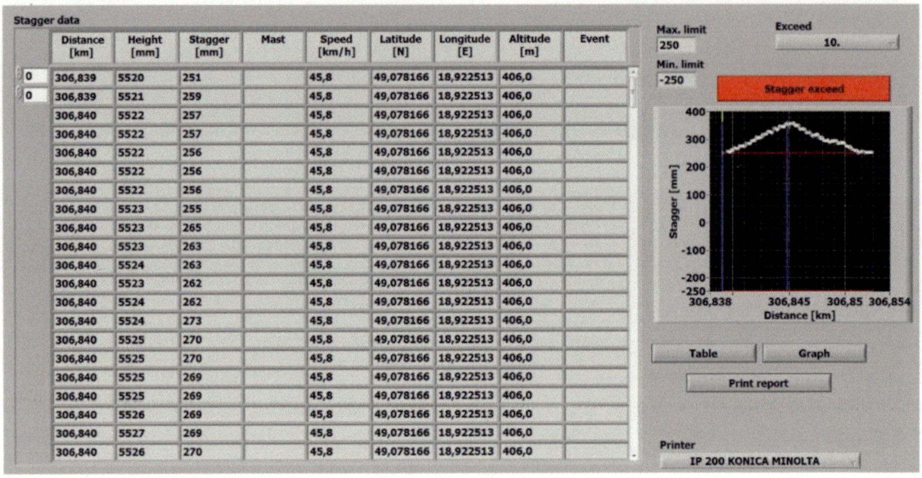

Figure 20.
Exceeding the maximum values.

Subsequently, all values that are greater than the specified limit values are automatically searched. These values can be displayed in a tabular or graphical form (**Figure 20**). Since it is very often not just individual values but a series of values, it is possible to select an individual series of values based on the analysis. The evaluated data can then be exported to a file or printed as a report.

Report creation is also possible for other measured parameters than just height and stagger exceedances. It is also possible to create a report containing data only on the value of height and stagger for specific masts. This report is important for checking the correctness of the overhead traction line placement.

The "Report Generation" toolkit is used to create the report. The user can select the choices on the application front panel, and based on their choices, the functions like "Append Image to Report" or "Append List to Report" are applied in the result. These functions are placed in the "Case" structure, and the user choice switches the "True" or "False" state (**Figure 21**).

When measuring the same section repeatedly, it is possible to compare the parameters measured during individual measurements (**Figure 22**). Sections must be selected from already evaluated files. For a correct evaluation, it is necessary to select the same measured sections. Synchronization of data for comparison is done based

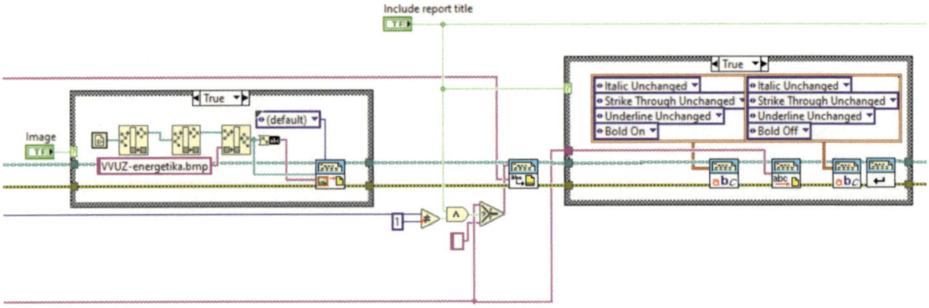

Figure 21.
The part of a block diagram for report generation.

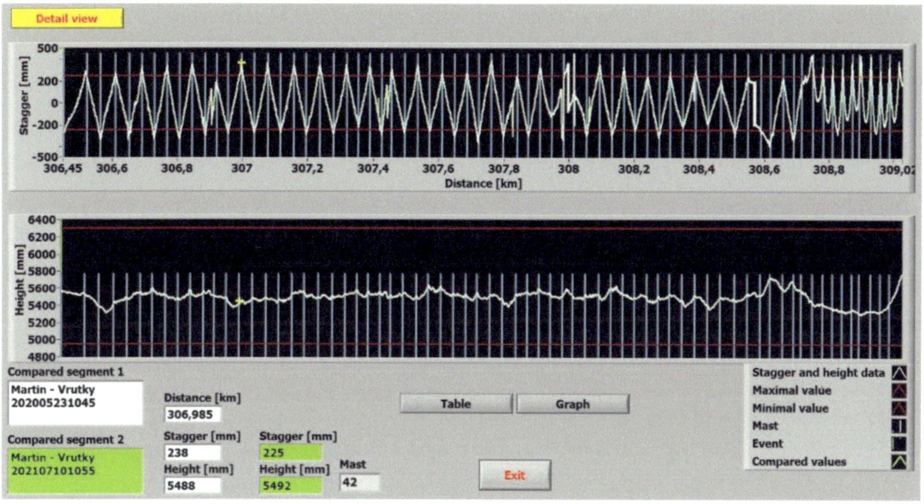

Figure 22.
Comparison of measured parameters for two measurements for the same section.

on the coordinates from the GPS module. In the basic graphic display, it is possible to view the measured data for the entire measured section, or it is also possible to choose a detailed display of the measured data. Measured data of height and stagger can also be determined using cursors located in graphs, while data from individual measurements are also represented numerically.

5. Conclusion

The presented device and developed software for overhead traction line height and stagger measurement are determined for continuous measurement of statistical parameters. The automatic data acquisition is the device's advantage, while the distance of measurement can be set by the user. The setting of measurement can be realized in the device through a keyboard or in a graphical user interface through a PC.

The device's internal memory is organized as a ring buffer, and measured data are recorded here. The recorded data must be evaluated through the developed software. This software enables evaluation based on limits set by technical norms.

Then, the measured parameters can be evaluated objectively because not only the absolute values of line height or stagger are presented but also the corrected values are displayed as real values.

The developed software can be extended by the possibility of an automated section measurement in the future. This would be possible based on the data from the GPS sensor, while the knowledge of the position of the masts at the beginning and end of the measured sections would be used.

Acknowledgements

Results of this research were supported by grant No. APVV 20-0500 - Research of methodologies to increase the quality and lifetime of hybrid power semiconductor modules and this publication was realized with support of Operational Program Integrated Infrastructure 2014 - 2020 of the project: Innovative Solutions for Propulsion, Power and Safety Components of Transport Vehicles, code ITMS 313011V334, co-financed by the European Regional Development Fund.

Conflict of interest

There is no conflict of interest.

Author details

Dušan Koniar*, Libor Hargaš, Peter Šindler and Marek Paškala
Faculty of Electrical Engineering and Information Technology, Department of Mechatronics and Electronics, University of Žilina, Žilina, Slovakia

*Address all correspondence to: dusan.koniar@feit.uniza.sk

References

[1] Hu B, Chan RWK. Railway overhead wiring structures in Australia: Review and structural assessment. Applied Sciences. 2022;**12**(3). Article 1492. DOI: 10.3390/app12031492

[2] Fararooy S, Mair C. Review of Railway Overhead Wire Geometry Measurement System, Current Collections for High-Speed Trains, IEEE Seminar, 2. Oct 1998. DOI: 10.1049/ic:19981003

[3] Hofler H, Dambacher M, Dimopoulos N, Jetter V. Monitoring and inspecting overhead wires and supporting structures. In: Proceedings of the IEEE Intelligent Vehicles Symposium, Parma, Italy, 14-17 June 2004; IEEE: Piscataway, NJ, USA, 2004

[4] Chen P, Xu C. Li Z, Xu Z. Design and implementation of non-contact detection system for catenary based on double linear array cameras. In: Proceedings of the 5th International Conference on Computer-Aided Design, Manufacturing, Modeling and Simulation, Busan, Korea, 22-23 April 2017

[5] Yu L, Gao S, Zhang D, Kang G, Zhan D, Roberts C. A survey on automatic inspections of overhead contact lines by computer vision. IEEE Transactions on Intelligent Transportation Systems. 2022;**23**(8). DOI: 10.1109/TITS.2021.3119023

[6] Aydin I, Karakose M, Akin E. Anomaly detection using a modified kernel-based tracking in the pantograph-catenary system. Expert Systems with Applications. 2015;**42**:938-948

[7] Hisa T, Kanaya M, Sakai M, Hamaoka K. Rail and contact line inspection technology for safe and reliable railway traffic. Hitachi Review. 2012;**61**:325-330

[8] Arastounia M. Automated recognition of railroad infrastructure in rural areas from LIDAR data. Remote Sensing. 2015;7:14916-14938

[9] Chen L, Xu C, Lin S, Li S, Tu X. A deep learning-based method for overhead contact system component recognition using mobile 2D LiDAR. Sensors. 2020;**20**:2224

[10] Harter M, Schipper T, Zwirello L, Ziroff A, Zwick T. Detection of overhead contact lines with a 2D-digital-beamforming radar system for automatic guidance of trolley trucks. International Journal of Vehicular Technology. 2013;**2013**:914351

[11] Harada S, Kusumi S. Monitoring of Overhead Contact Line Based on Contact Force. In: IET International Conference On Railway Condition Monitoring, November 2006, Birmingham, UK, ISBN: 0-86341-732-9

[12] Ostlund S, Gustafsson A, Buhrkall L, Skoglund M. Condition monitoring of pantograph contact strip, 4th IET International Conference on Railway Condition Monitoring, June 2008, Derby, UK, ISSN: 0537-9989, DOI: 10.1049/ic:20080343

[13] Shing AWC, Pascoschi G. Contact Wire Wear Measurement and Data Management. In: IET International Conference On Railway Condition Monitoring, November 2006, Birmingham, UK, ISBN: 0-86341-732-9

[14] NXP Semiconductors. LPC2212/2214, Single-chip 16/32-bit ARM microcontrollers; 128/256 kB, ISP/IAP flash with 10-bit ADC and external memory interface (Product datasheet), Rev. 5-14 June 2011. Available from:

https://www.alldatasheet.com/datasheet-pdf/pdf/222015/PHILIPS/LPC2214.html

[15] OPTEK Technologies. Long Distance Reflective Switch OPB732, OPB732WZ (Product datasheet). Available from: https://pdf1.alldatasheet.com/datasheet-pdf/view/139962/OPTEK/OPB732.html

Chapter 5

Virtual Instrumentation Used in Renewable Energy

Petru Adrian Cotfas, Daniel Tudor Cotfas and Horia Hedesiu

Abstract

The demand of energy increases once with the growth of population, and therefore, the finding of or improvement in the efficiency of renewable energy sources becomes very important for researchers and industry. The conversion of solar energy into electrical energy can be done based on photovoltaic or Seebeck effects. In the first case, photovoltaic panels are used, while in the second case, thermoelectric generators are used. The two can be combined to obtain the so-called hybrid systems, which have the goal to improve the overall conversion efficiency of the system. This chapter is focused on showing how the graphical programing language, called NI LabVIEW, together with a SPICE simulator, called NI Multisim, can be used for studying and understanding the behavior of the photovoltaic and thermoelectric generators as parts of the renewable energy sources. Different simulations developed in LabVIEW or Multisim are presented, and some monitoring and characterization applications are also described. Simple simulations to complex laboratory or industrial-level applications are dealt with in this chapter.

Keywords: virtual instrumentation, LabVIEW, simulations, real-time and FPGA platforms, photovoltaic cells, thermoelectric generator

1. Introduction

According to a recent analysis [1], the world's population reached 8 billion people in November 2022. Therefore, the energy demand of the entire population is increasing rapidly. The actual used energy is obtained from different energy sources (ES), but unfortunately, the largest percent of these ES is based on fossil resources, which are very polluting and limited. In order to have a sustainable civilization from the energy point of view, renewable energy sources (RES) should be used. RES include solar energy, wind energy, geothermal energy, biomass-based energy, waves energy, and so on.

One problem of using RES is the variability of energy availability in time. For example, in the solar energy case, the availability depends on the moment of time – day or night – and the climate conditions – cloudy or clear sky. Therefore, the usage of this RES must be combined with another or implemented with energy storage solutions – like batteries. Such solutions are called hybrid solutions. The solar energy is converted into electrical energy and thermal energy. In the first case, the most used conversion technologies are based on the photovoltaic and Seebeck effects. The photovoltaic effect is exploited using photovoltaic cells and panels (PV). The Seebeck effect is exploited using thermoelectric generators (TEG). The conversion of solar energy into thermal energy is done by using so-called solar thermal collectors, which absorb the solar radiation and heat the internal thermal agent, which can be water, air, nanofluid, or oil [2]. The combination of these energy conversion systems is used nowadays more often in order to improve the overall efficiency of the system. There are different hybrid systems for solar energy conversion, such as [3]:

- Photovoltaic-thermal (PV-T) system that uses the PV to convert the solar irradiance into electricity and solar collector to heat the thermal agent but also to cool down the PV in order to increase the PV efficiency and its life time.

- Photovoltaic-thermoelectric generator (PV-TEG) system that uses both components to generate electricity. The PV absorb around 95% of the solar irradiance and convert, depending on the PV type, between 10 and 25% into electricity, while the rest is converted into heat, which increases its temperature and decreases its efficiency. The TEG based on the Seebeck effect absorbs the heat from the PV and generates electricity.

- Photovoltaic-thermoelectric generator-thermal (PV-TEG-T) system that combines all three components to generate electrical and thermal energy.

In order to study or monitor RES-based systems, different software and hardware should be used. There are many solutions in the market; some of them are dedicated to a specific RES, while others are dedicated to the general use.

This chapter is focused on using the NI LabVIEW graphical programming languages for RES study. The NI LabVIEW is used at all levels: simulation level, measurement and characterization level, and monitoring and industrial level.

NI LabVIEW is a development software platform that allows the implementation of the virtual instrumentation (VI) concept. According to NI, VI represents “a combination of modular hardware and customizable software used to create measurement and control system defined by user.” In fact, a virtual instrument is a measurement and control instrument based on a computing system (in the majority of cases, it is based on a PC). The VI is an alternative to the traditional instrumentation (TI). The TI normally has a functionality and user interface defined by the manufacturer. Instead, the VI functionality and user interface are defined by the user and can be easily updated or replaced.

Considering the advantages offered by the VI and NI LabVIEW, these are used in this chapter to develop simulated and real instruments for RES study, focused on PV systems and some hybridization solutions.

The chapter is split into four sections, as follows: in the second section is discussed the implementation of the simulations for PV and TEG components based on different tools. The third section is dedicated to the implementation from simple to complex methods for PV and TEG characterization. The fourth section shows how a monitoring system based on a real-time (RT) platform for a home PV system is implemented. This section will show how to use VI in Industrial Internet of Things (IIoT) implementation.

2. Simulation

This section is dedicated to show how to use the NI LabVIEW for PV and TEG study through simulations.

2.1 Theoretical aspects: PV

The PV are used for converting the solar irradiance to electricity. A PV panel is composed of a set of PV cells combined in a series and/or parallel connection in order to obtain the desired voltage and current. The most used model of a PV cell is the so-called one diode model, which is shown in **Figure 1**.

The diagonal cross fill from **Figure 1** marks the one diode model of the PV whose mathematical description is given by eq. (1).

$$I = I_{ph} - I_0 \left(e^{\frac{q(V+IR_s)}{nk_B T}} - 1 \right) - \frac{V + IR_s}{R_{sh}} \tag{1}$$

where V and I are the output voltage and current of the PV cell, I_{ph} represents the photogenerated current, I_0 is the reverse saturation current, q is the elementary electrical charge, R_S is the series resistance of the PV cell, R_{SH} is the shunt resistances of PV cell, n is the ideality factor of the diode D, k_B is the Boltzmann constant, and T is the PV cell temperature. The quantity V_T (2) is called thermal voltage of the diode and has a value ~26 of mV at room temperature.

$$V_T = \frac{k_B T}{q} \tag{2}$$

By connecting N_S PV cells in a series, a PV panel can be obtained. The PV panel's mathematical description is given by eq. (3).

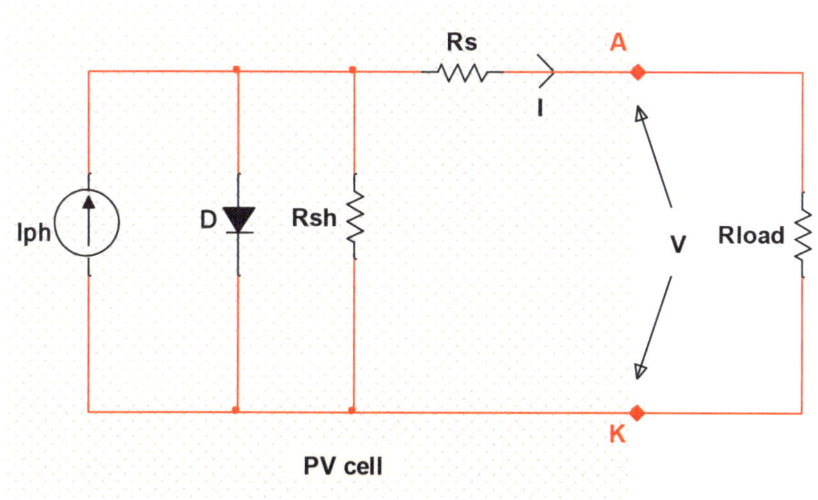

Figure 1.
The one diode model of a PV cell.

$$I = I_{ph} - I_0\left(e^{\frac{q(V+N_sIR_s)}{N_Snk_BT}} - 1\right) - \frac{V + N_sIR_s}{R_{sh}} \tag{3}$$

where all parameters refer now to a PV panel and not to a PV cell.

Considering the PV panel parameters offered by the manufacturer, such as: V_{OC} – open circuit voltage, I_{SC} – short circuit current, k_V and k_I – current and voltage thermal coefficients of the PV panel, R_{SH}, R_S, n, and N_S, the current-voltage (I-V) characteristics of the PV panel can be determined for different levels of irradiance and temperature using the following equations:

$$V_{OC}(T) = V_{OC}(T_{REF}) + k_V(T - T_{REF}) \tag{4}$$

$$I_{SC}(T) = I_{SC}(T_{REF})\left(1 + \frac{k_V}{100}(T - T_{REF})\right) \tag{5}$$

$$I_0(T) = \left(I_{SC}(T) - \frac{V_{OC}(T) - I_{SC}(T)R_S}{R_{SH}}\right)e^{-\frac{V_{OC}(T)}{N_SV_T}} \tag{6}$$

$$I_{PH}(T) = I_0(T)e^{\frac{V_{OC}(T)}{N_SV_T}} + \frac{V_{OC}(T)}{R_{SH}} \tag{7}$$

$$I_{SC}(G) = I_{SC}(T)\frac{G}{1000} \tag{8}$$

$$I_{PH}(G) = I_{PH}(T)\frac{G}{1000} \tag{9}$$

$$V_{OC}(G) = ln\left(\frac{I_{PH}(G)R_{SH} - V_{OC}(T)}{I_0(T)R_{SH}}\right)N_SV_T \tag{10}$$

$$I(G,T) = I_{PH}(G) - I_0(T)\left(e^{\frac{(V+I(G,T)R_S)}{N_SV_T}} - 1\right) - \frac{V + I(G,T)R_S}{R_{SH}} \tag{11}$$

$$D_V = V_{OCi}(G) - V_{OCi-1}(G) \tag{12}$$

$$D_I = I_i(G,T) - I_{i-1}(G,T) \tag{13}$$

where T_{REF} is the reference temperature (25°C), and G is the irradiance level. Eqs. (10) and (11) are implicit ones and therefore are calculated iteratively. The stop conditions for the iterations are achieved by the predefined maximum number of iterations or the difference between values obtained at two consecutive iterations D_V and D_I. The calculus of these differences is based on eqs. (12) and (13), respectively.

2.2 Theoretical aspects: TEG

A TEG is a device that allows the conversion of thermal energy into electrical energy. It is based on the Seebeck effect, which consists of an electromotive force that appears at the terminals of an electrical circuit made up of two different types of materials, when there is a temperature difference between the terminals. A TEG consists of a set of thermo-elements (modules or uni-couples) connected in series. A thermo-element is formed by two legs: a p-type and an n-type semiconductor, connected in series through a metallic conductor (copper), as shown in **Figure 2a**. The simple equivalent electric circuit of a TEG consists of a thermal dependent voltage source connected in series with a resistor (**Figure 2b**).

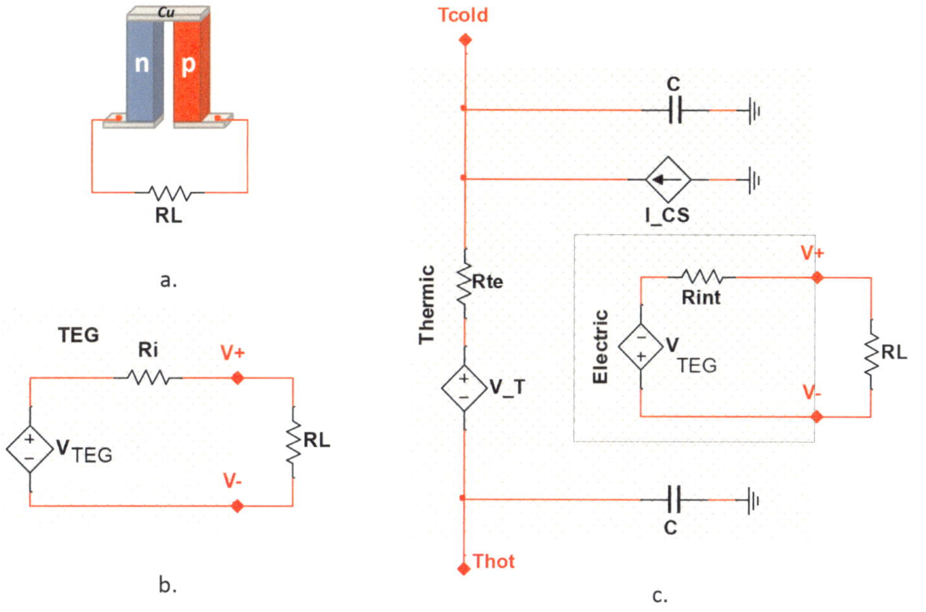

Figure 2.
The model of a TEG. a. the TEG pellet; b. simple model of a TEG; and c. thermal and electrical model of a TEG.

In order to model the thermal dependence of the voltage source, the equivalent circuit must be updated as shown in **Figure 2c**, which is described by the following equation:

$$V\_T = IR_{te}\left(\alpha_{Sb}T_{hot} - \frac{IR_{int}}{2}\right) \tag{14}$$

$$V_{TEG} = \alpha_{Sb}(T_{hot} - T_{cold}) \tag{15}$$

$$I\_CS = V \cdot I \tag{16}$$

where V_T is a voltage source that models the Peltier cooling and heating on the cold and hot sides, R_{te} is the thermal resistance of the TEG, α_{Sb} is the Seebeck coefficient, R_{int} is the electrical internal resistance of the TEG, V_{TEG} is the voltage source that corresponds to the Seebeck effect, I_CS is the current source that corresponds to the Joule heating of the TEG, I is the output current generated by the TEG, V is the output voltage of the TEG, T_{hot} and T_{cold} are the temperatures of the hot and cold sides of the TEG, *and* C is a capacitor that models the lumped heat capacitance of the ceramic plates of the TEG.

2.3 LabVIEW simulation

A possible LabVIEW implementation of the PV panel simulation is shown in **Figure 3**. **Figure 3a** represents the user interface of the main LabVIEW application that allows users to introduce the PV panel parameters, **Param In**; the level of irradiance, **G**; and the PV working temperature, **T**. The user can also introduce the starting voltage and voltage step for the I-V characteristics calculus. On the two graphs of the

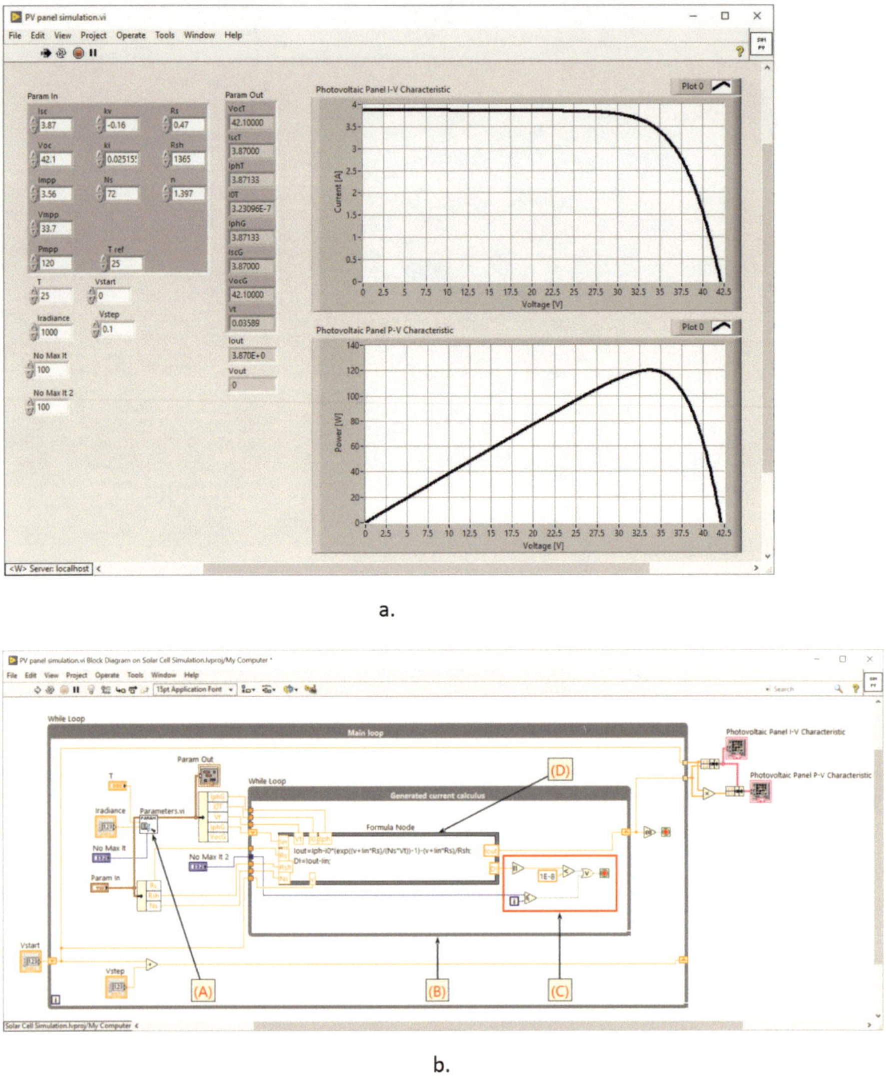

Figure 3.
LabVIEW implementation of the PV panel simulation.

application are shown the I-V and power-voltage (P-V) characteristics of the PV panel. The application also displays the PV panel parameters for the set temperature and irradiance level, **Param Out**.

Figure 3b presents the LabVIEW code for PV panel simulation. The Parameters.vi (A) is a subVI used to determine the PV parameters at the desired condition (T and G) based on eqs. (4)–(10) and (2) (see **Figure 4a** and **b**). The generated current, I, is calculated using the *Formula Node* structure (D). The current is determined iteratively using the *While Loop* structure (B), which uses D_I as finish conditions whose value should be smaller than 10^{-8} or the maximum number of iteration (C). The same implementation approach is used to determine the value of V_{OC} for the desired T and G (**Figure 4a**). In **Figure 4b**, the value of V_T is determined using eq. (2).

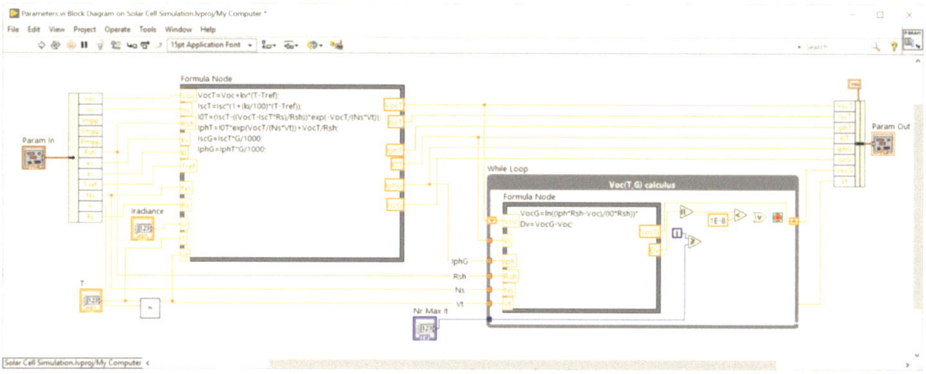

a.

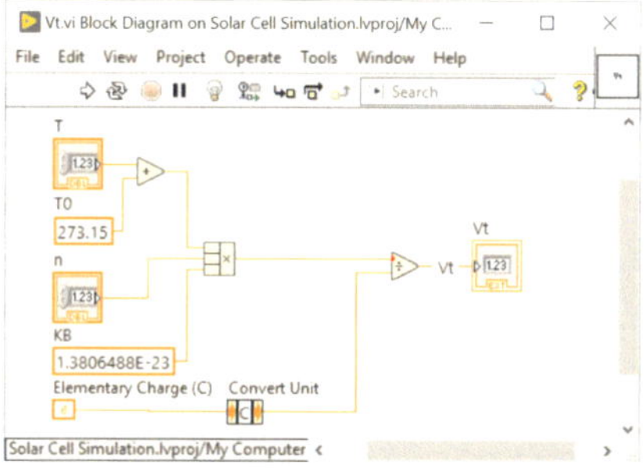

b.

Figure 4.
PV panel parameters calculus.

2.4 LabVIEW and NI Multism simulation

2.4.1 PV simulation

An interesting option offered by LabVIEW is to use the NI LabVIEW Multisim API Toolkit that allows to interconnect the LabVIEW programming languages with a SPICE simulator. NI Multisim is an application for SPICE simulation and circuit design [4] and together NI Ultiboard represents a complete software package for circuit design, simulation, validation, and PCB layout.

If the LabVIEW Multisim API Toolkit is installed, the corresponding library is available in the function palette of LabVIEW (**Figure 5**) at the following path *Function> > Connectivity> > Multisim*.

Developed models of PV cells and PV panels in NI Multisim using SPICE code are described in the study case “New Models for Photovoltaic Cells in Multisim” published on the NI website (see [5]). The PV panel model developed is based on eqs. (2)–(11) (**Figure 6**). The PV model is encapsulated in the **U1 PV** component.

Based on the *Parameter Sweep* analysis, the I-V characteristic can be easily obtained in Multisim. The used parameters of the analyses are shown in **Figure 7a**. The output parameter is set in the *Output* tab as the current through the V_{bias} source. This source is a voltage source for PV panel polarization. V_{irrad} is used to fix the level of irradiance. A value of 1000 V means an irradiance level of 1000 W/m^2 (1 sun). **Figure 7b** shows the obtained I-V characteristics of the PV panel.

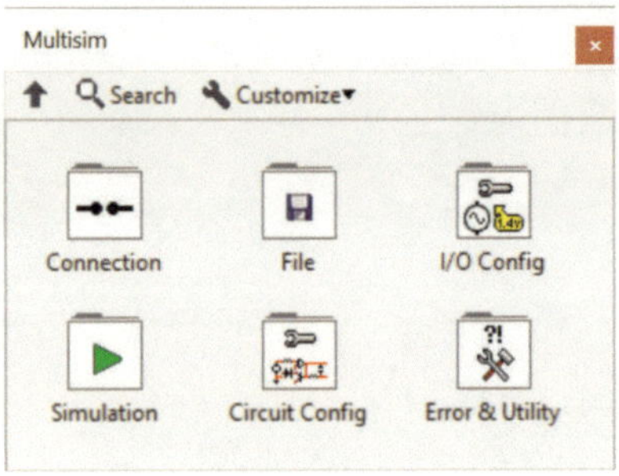

Figure 5.
The LabVIEW Multisim palette.

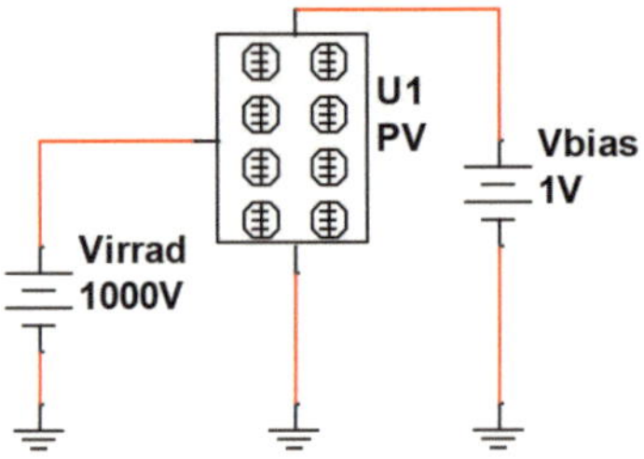

Figure 6.
The Multisim model of a PV panel.

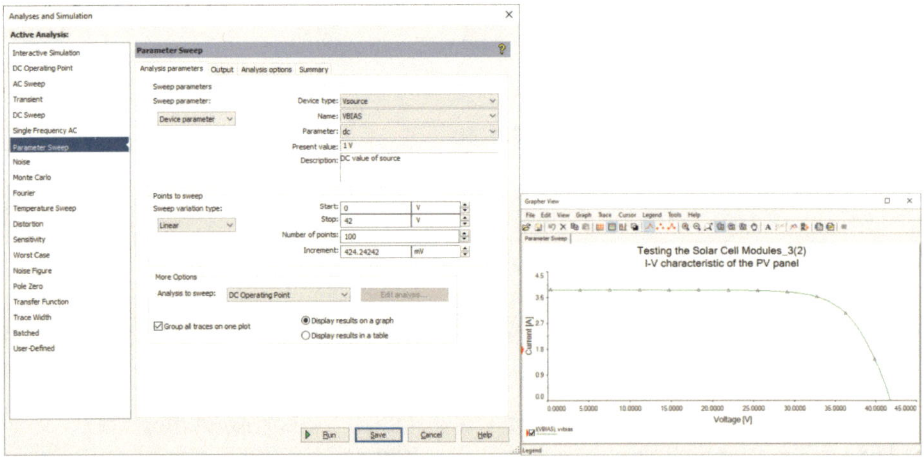

a. b.

Figure 7.
The Multisim model of a PV panel.

The Multisim allows combining the variation of more parameters during the same analyses, so that the effect of the irradiance or temperature variation could be easily studied. In **Figure 8**, the Parameter Seep setup for studying the influence of the irradiance level over the PV panel response is shown. The value of the V_{irrad} source is used as the first sweep parameter. The range of variation is fixed between 400 and 1000 with a step of 200. In the *More option* field, the *Nested sweep* option is chosen (**Figure 8a**). By pressing the *Edit analysis* button, the second sweep parameter is set

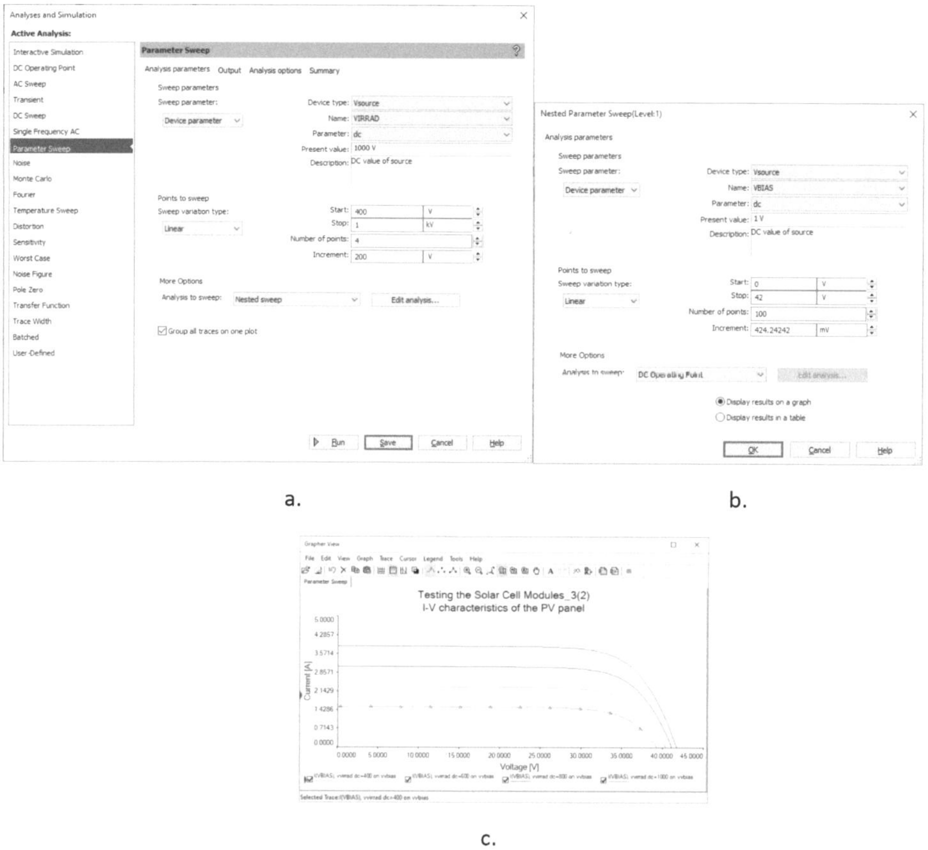

Figure 8.
The influence of the irradiance level on the PV panel response using nested sweep analysis.

(**Figure 8b**). The range value of the V_{bias} source is fixed between 0 and 42 volts with 100 equidistant number of points. The *Analysis to sweep* option is fixed to the *DC Operating Point* value. The obtained results are shown in **Figure 8** c.

These analyses can be performed directly from LabVIEW by using the NI LabVIEW Multisim API Toolkit. The advantage of such an approach is the fact that the data are obtained directly in the LabVIEW environment and can be easily compared to the data obtained from real experiments.

In order to be able to call the Multisim circuit directly from LabVIEW, it is necessary to add desired probes in the circuit (**Figure 9**). These probes become circuit outputs, which can be read in the LabVIEW application. **Figure 9** shows the PV panel circuit modified by adding two probes for current, $\boldsymbol{I_{out}}$, and voltage, $\boldsymbol{V_{out}}$, reading.

Based on Multisim API subVIs, a LabVIEW application was developed to interrogate the Multisim circuit. In **Figure 10**, the panel and diagram of the application are presented. The *Connect.vi* sets the connection between LabVIEW and Multisim; then, the circuit is opened using *Open File.vi*, which has the path of the circuit file as input. *Set Input Data.vi* and *Ramp Pattern.vi* are used to set the values of the $\boldsymbol{V_{bias}}$ voltage source. *Enum Outputs.vi* allows to find all probes defined in the circuit. In this example, all probes are read and therefore are passed to the *Set Output Request.vi* in order to have the values of the voltage and current determined through simulation directly in LabVIEW. After the circuit is configured, the simulation is started with *Run*

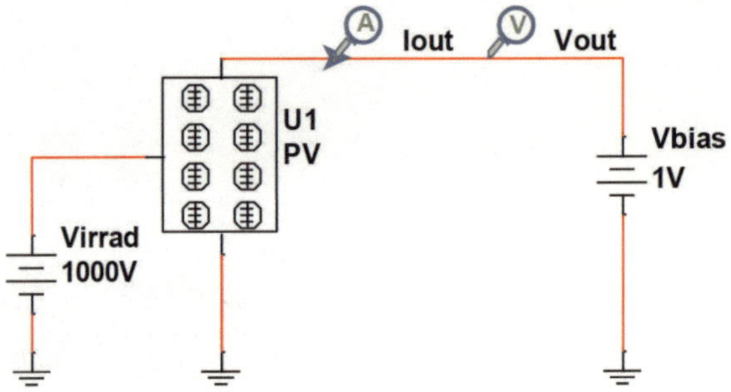

Figure 9.
The modified Multisim circuit with current and voltage probes.

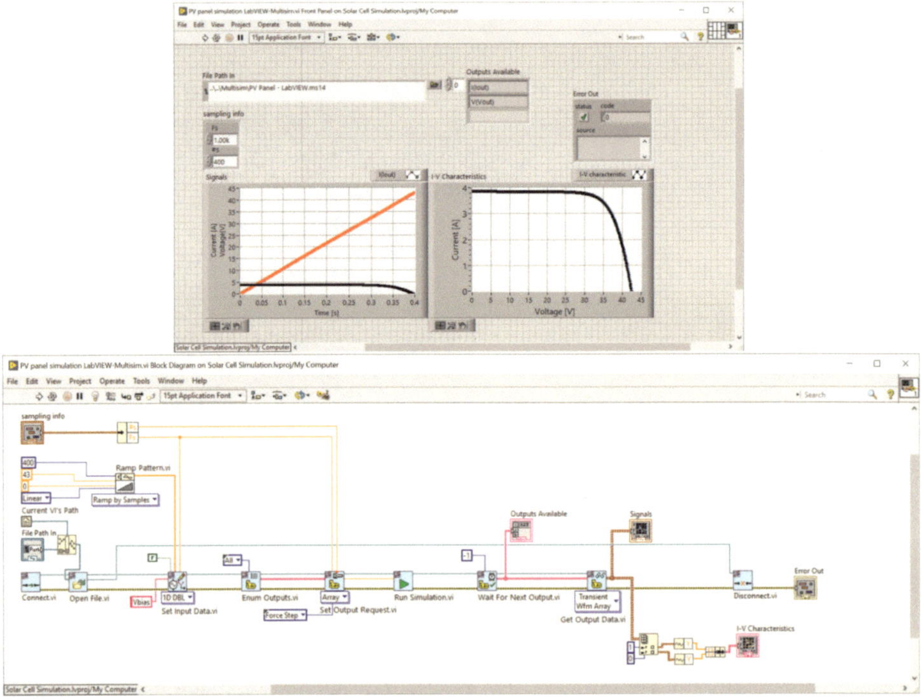

Figure 10.
The LabVIEW application developed with NI LabVIEW Multisim API.

Simulation.vi, and with *Wait for Next Output.vi*, the simulation ending is monitored. *Get Output Data.vi* returns data obtained through simulation, which are displayed on a Waveform Graph as individual signals (current and voltage generated by the PV panel) and as I-V characteristics. *Disconect.vi* is used to close the connection between LabVIEW and Multisim.

A very important parameter that considerably influences the PV panel efficiency is the temperature of the PV panel. Therefore, by adding the temperature parameter in the Multisim circuit (**Figure 11**), its influence on the PV panel efficiency can be easily investigated.

The LabVIEW application should now include the *Set Circuit Parmeters.vi*, which allows modifying the TEMP parameters and through it the PV panel behavior (see **Figure 12**). By introducing the simulation code in a *For Loop* and using the *Simulation*

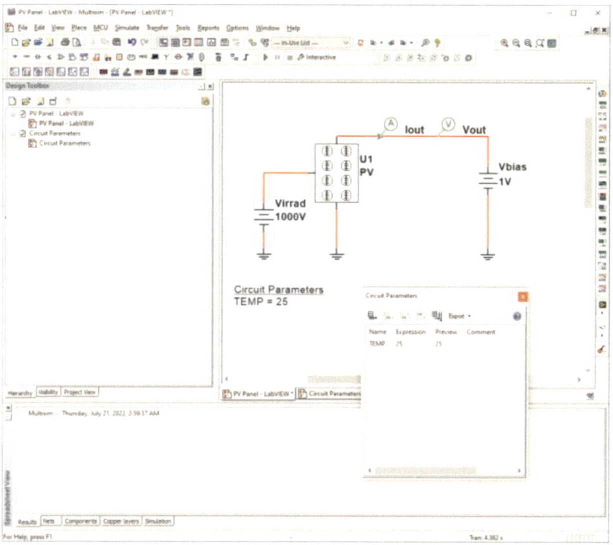

Figure 11.
Activating the TEMP circuit parameter in Multisim.

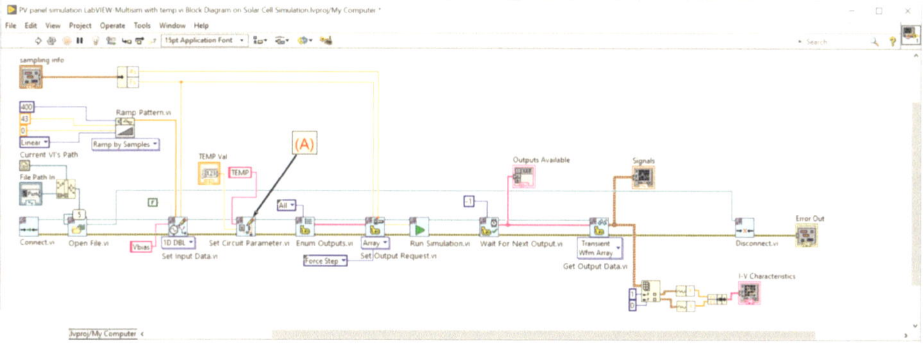

Figure 12.
The LabVIEW application with TEMP circuit parameter.

State.vi and *Simulation Stop.vi,* as shown in **Figure 13**, the effect of temperature over the PV panel response can be visualized (see the graph in **Figure 13**).

2.4.2 TEG simulation

The same approach as in the case of PV panels can be applied in the case of TEGs. The TEG model developed using the SPICE code is presented in [6] and is based on eqs. (14)–(16). The TEG model is encapsulated in the **U2 TEG** component (**Figure 14**).

The T_{hot} and T_{amb} DC sources model the temperatures of the TEG sides, cold and hot. The V_{pol} is a voltage source used for biasing the TEG. Using the Parameters Swipe analysis, the I-V characteristics of TEG at different temperatures can be obtained. The Parameters Sweep setup with the Nested Sweep option is shown in **Figure 15a** and **b**. **Figure 15c** shows the obtained I-V characteristics of the TEG, for the hot side

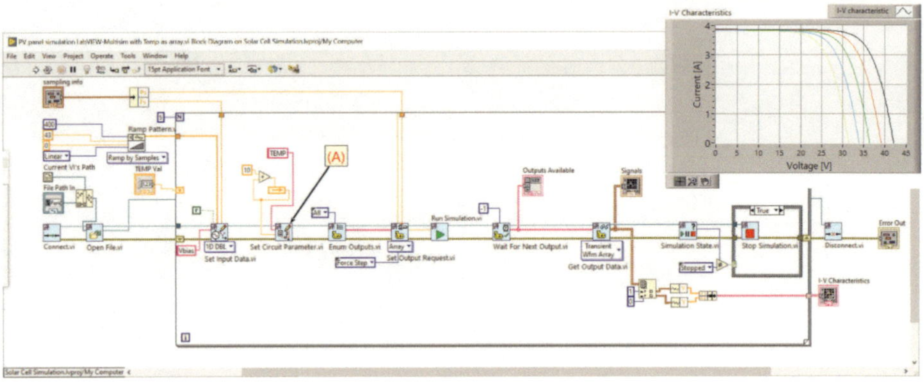

Figure 13.
The LabVIEW application with multiple values of TEMP circuit parameter.

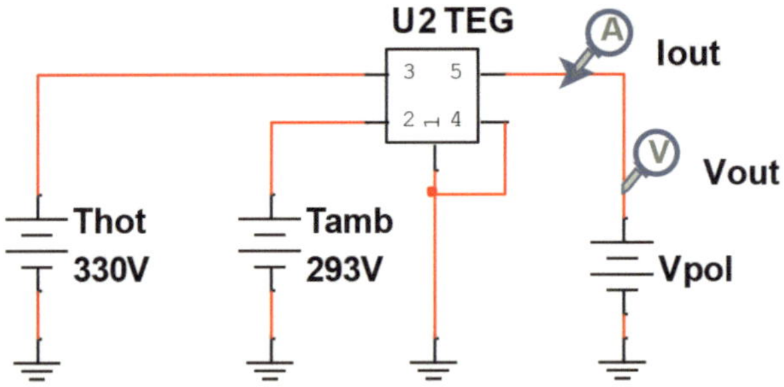

Figure 14.
The Multisim model of a TEG panel.

temperature varying between 300 and 330 K, considering that the cold side temperature is constant at 293 K.

Using the LabVIEW application shown in **Figure 10**, adapted to the new schemata, the I-V characteristic can be obtained. The modification consists of changing the name of the used source from V_{bias} to V_{pola1} and the range of the signal applied from 0 to 42 V to 0–0.4 V, as can be seen in **Figure 16**. The temperature for T_{hot} was set at a value of 330 K.

2.5 Control and simulation toolkit

Another option for implementing the simulation of PV systems is to use the LabVIEW Control Design and Simulation (CDS) toolkit. This toolkit allows to develop, implement, and analyze the behavior of different kinds of dynamic systems. In this chapter, we focus on using the simulation part of this toolkit applied in PV simulation [7]. **Figure 17** presents the simulation palette that includes libraries like Signal Generation, Signal Arithmetic, Continuous Linear Systems, Nonlinear Systems, Discrete Linear Systems, etc. One particular library is dedicated to accessing external models defined on other software and saved as shared library – dll or accessing Multisim models. Next, the use of the second option is presented. The used model is a modified model of the one presented in **Figure 6**. In this model (**Figure 18**), we used a voltage-controlled source ***V1*** as the PV panel biasing source. The ***Irad***, ***Vbias***, ***Iout,*** and ***Vout*** are hierarchical connectors (the first two are inputs and the last two are outputs).

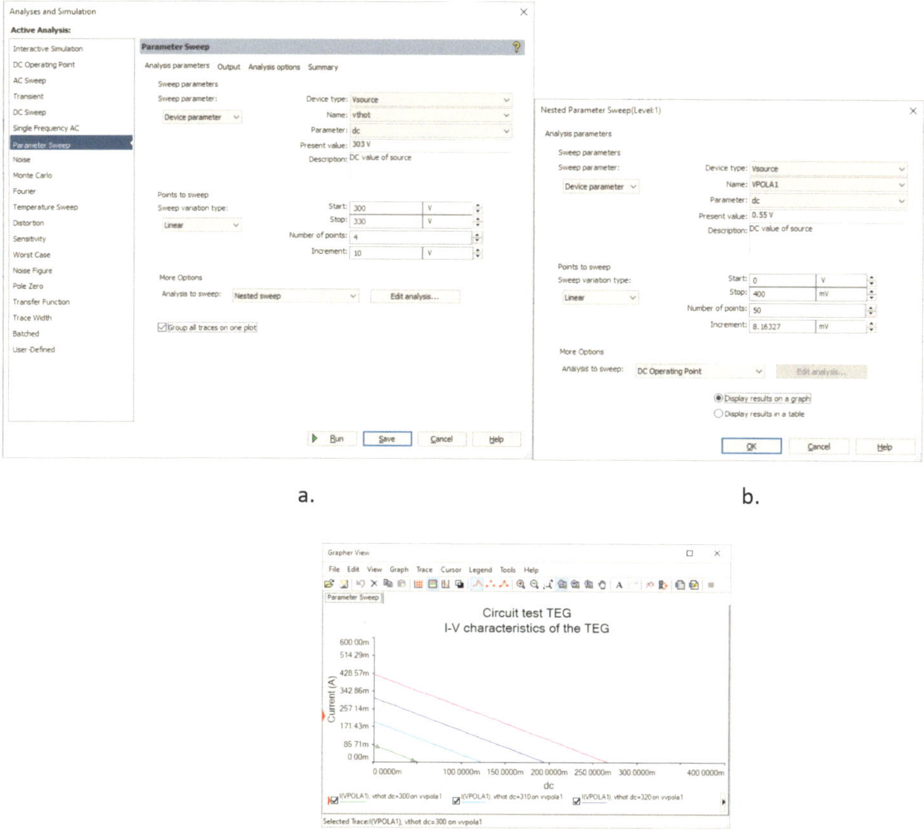

a. b.

c.

Figure 15.
The influence of temperature on the TEG response using the nested sweep analysis.

The first input is used for defining the irradiance incident on the PV, while the second one is used for controlling the ***V1*** source.

In order to read the PV current, a current clamp (XCP1) was used, with 1 mV/1 mA conversion ratio. The hierarchical connectors were used as model inputs and outputs in the LabVIEW application.

In order to develop a simulation application in LabVIEW, the *control & simulation loop* (see the red rectangle in **Figure 17**) is mandatory to be defined. The simulation functions can be used only inside of this loop.

A very simple application for PV simulation was developed in LabVIEW using the modified PV model (**Figure 18**) and the simulation functions of the CDS toolkit. The front panel and block diagram are shown in **Figure 19**. First, the *control & simulation loop* is set. Then, the *Multisim Design* node is used for Multisim model accessing. This node can be found at the following path: *Functions> > Control and Simulation> > Simulations> > External Models> > Multisim*. The node requires the path for the Multisim file. This node recognizes automatically the defined inputs and outputs in the Multisim file.

Therefore, on the ***Irrad*** input, a control is connected in order to have the possibility to modify the irradiance level incident on the PV panel. To control the PV bias, a *Ramp Signal* node (*Functions> > Control and Simulation> > Simulations> > Signal*

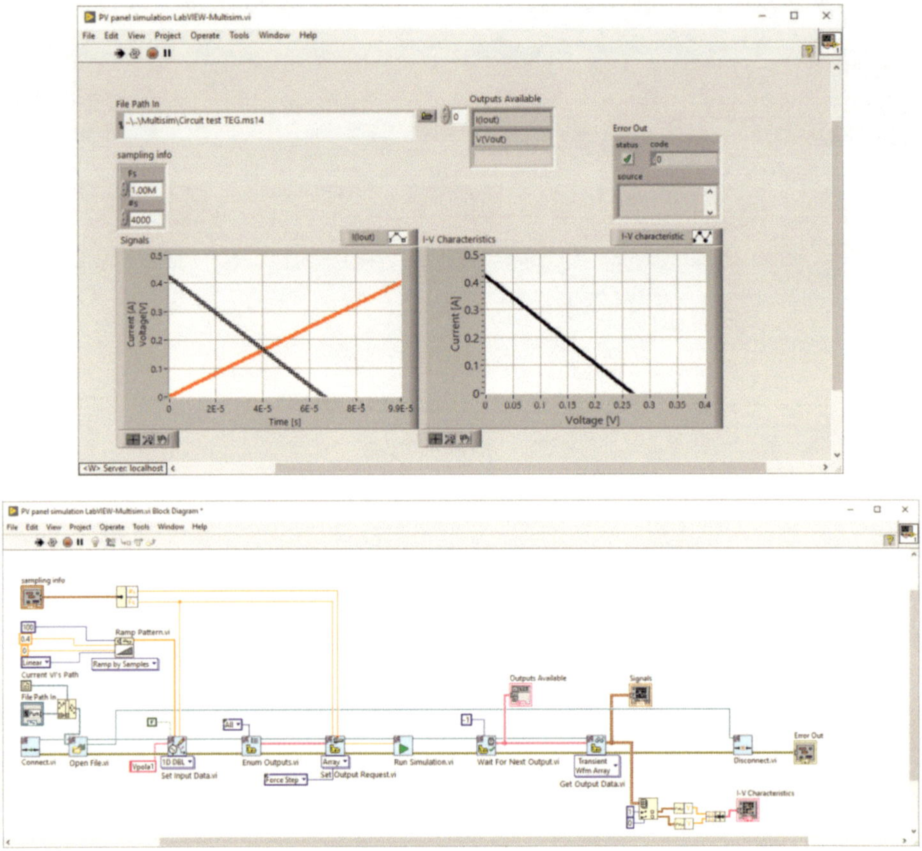

Figure 16.
The front panel and bloc diagram of the LabVIEW application for TEG.

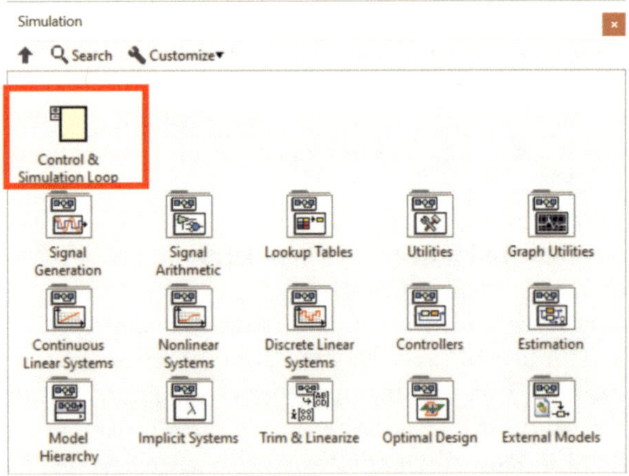

Figure 17.
The simulation palette of LabVIEW control design and simulation toolkit.

Generation) is defined and configured so that its ***Slope*** parameter is available as a terminal and starts with the initial value fixed as zero. The output of this signal generator is connected to the ***Vbias*** input. The I-V and P-V curves are built based on

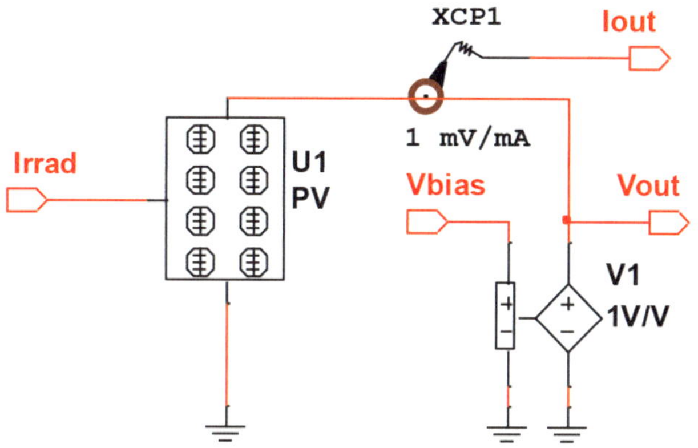

Figure 18.
The modified PV model developed in Multisim.

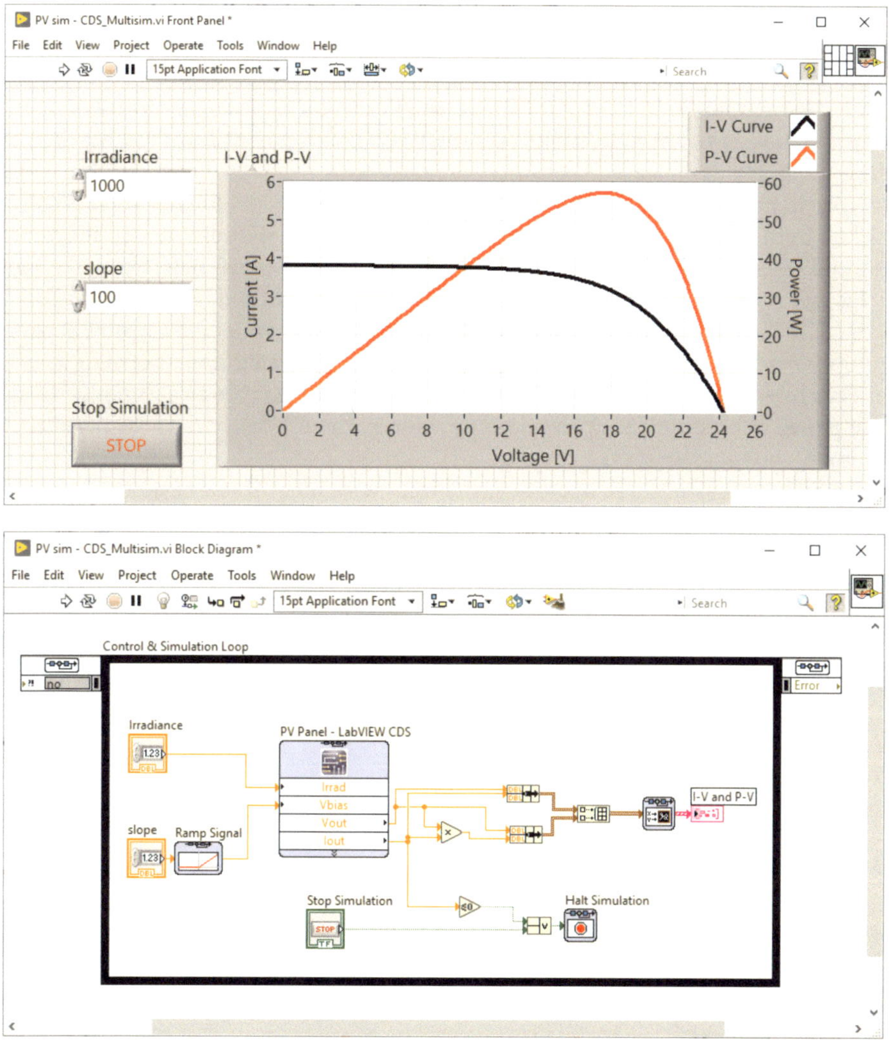

Figure 19.
The LabVIEW application developed with the CDS toolkit.

two *Bundle* functions, unified as an array with the *Build Array* function and then displayed on a *Buffered XY Graph* (*Functions> > Control and Simulation> > Simulations> > Graph Utilities*). The simulation is ended if the output current becomes less than or equal to zero or if the ***Stop Simulation*** button is pressed. For stopping the simulation loop, the *Halt Simulation* node (*Functions> > Control and Simulation> > Simulations> > Utilities*) is used.

3. Measurements and characterization

3.1 LabVIEW and simple I-V characteristics measurements

In this section, simple to complex methods for measurement of the I-V characteristics of PV panels are presented. There are many methods used for I-V characteristics measurements [8]. The simplest methods for I-V characteristics measurements are based on using dynamic electronic loads built based on a MOSFET transistor or a capacitor. The circuit schematic of dynamic loads based on the two components are shown in **Figure 20** (a shows the schematic for the MOSFET method, while b shows the capacitor method). In the MOSFET method case, by applying a ramp voltage between 0 and 5 V to the gate of the MOSFET (*Q1*), this passes form the blocked state to the completely open state. Therefore, the transistor behaves like a variable resistor, whose resistance varies from very high values to very low values. This variable resistor is applied to the PV panel, which passes from the open circuit point to the short circuit point, and by measuring the current and voltage at the PV terminals, the I-V characteristics can be determined. In the case of the capacitor method, through the SPDT switch (***S1***), the completely discharged capacitor (***C1***) is connected to the PV panel (***S1*** is on position 1). During the charging process, the internal impedance of ***C1*** varies from very low values to very high values and therefore behaves as a dynamic load for the PV panel. When the ***S1*** is passed on position 2, the ***C1*** discharges through the resistor ***R6***.

For a better measurement accuracy, the four-wire measurement method should be used, which is highlighted with thick blue circles in **Figure 20b**. For the current measurement, a power resistor with well-known resistance is used (***Rsense***). Measuring the voltage drop across the resistor and applying the Ohm law, the current value can be determined.

The resistors ***R2–R5*** are used as bias resistors for the differential setup of the DAQ device. The channels of the DAQ device used are as follows: ***AO0*** is the first analog

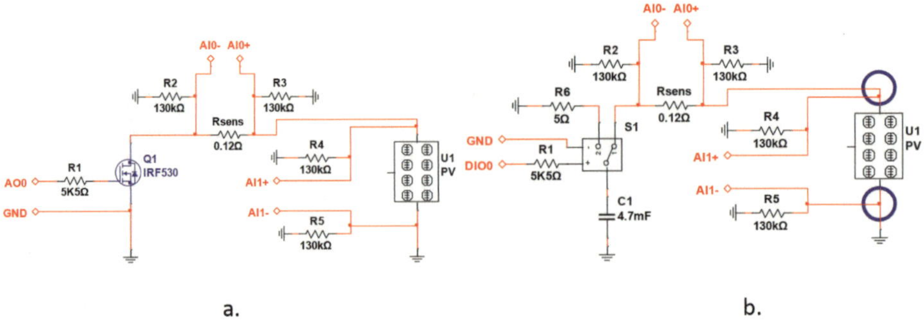

Figure 20.
The circuits schemata for I-V measurement. a. MOSFET method, b. capacitor method.

output channel used for generating a ramp voltage signal applied to the MOSFET gate, ***AI0- – AI0+*** and ***AI1- – AI1+*** are the first and second analog input channels configured in differential connection modes and are used for measuring the current and voltage generated by the PV panel. ***DIO0*** is the first digital line of the board configured as the output and is used to control the ***S1*** switch.

Based on the MOSFET circuit schemata, a PCB was designed and build as shown in **Figure 21**, and the entire system for a PV panel characterization is shown in **Figure 22**. The used DAQ board is NI USB 6215.

A very simple software application was developed based on the DAQmx library (**Figure 23**). The block diagram of the application is shown in **Figure 23a** and contains four zones. The first zone, (1), is dedicated to the analog signal generation used for MOSFET gate control. It contains the standard line of DAQmx subVIs for finite samples generation. The second zone, (2), is dedicated to the analog input measurement used for obtaining the current and voltage generated by the PV panel. It contains the standard line of DAQmx subVIs for finite samples acquisition on multiple channels. The third zone, (3), is used for building the signal that controls the MOSFET gate. A *Triangle Waveform.vi* was used and configured for a quarter or half period

Figure 21.
The electronic circuit based on MOSFET.

Figure 22.
The system for I-V characteristic measurement of a PV panel.

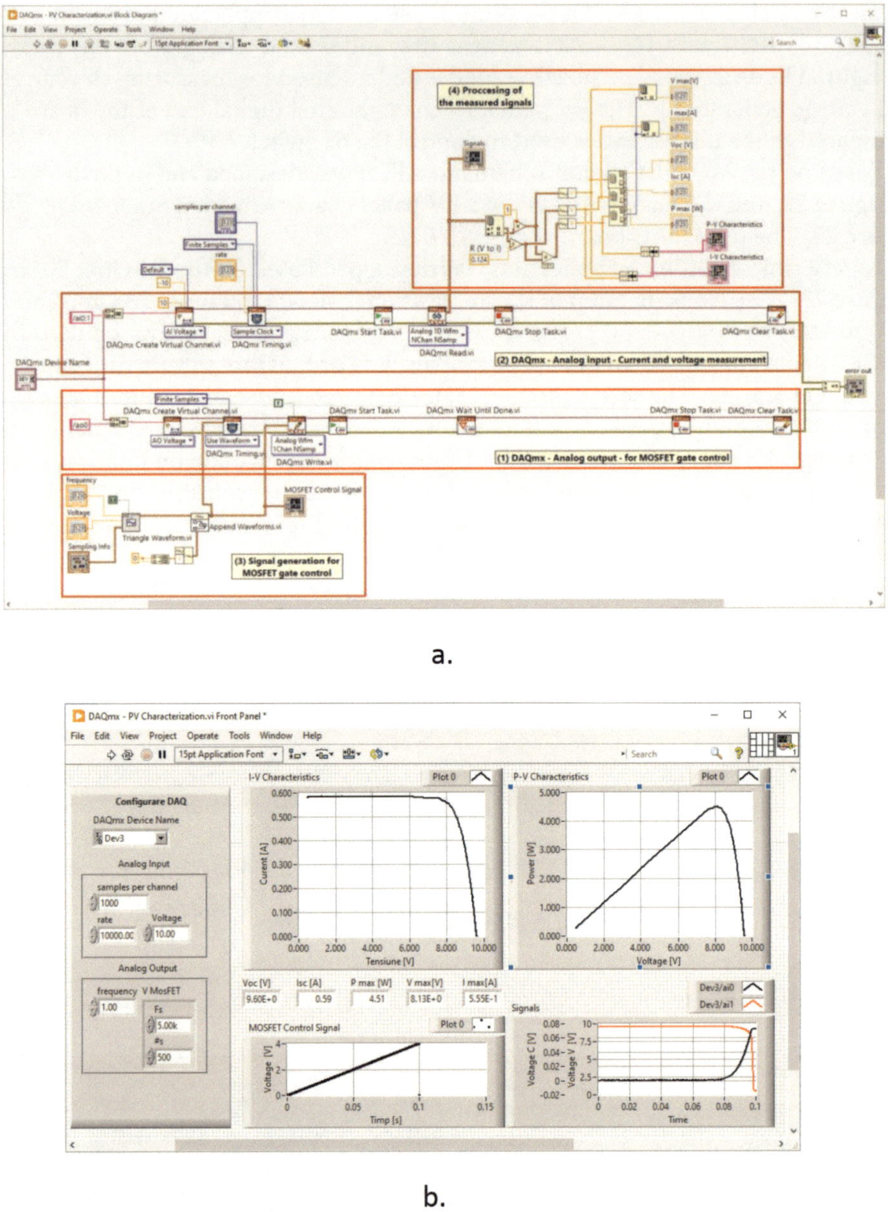

Figure 23.
The application for PV panel characterization based on the MOSFET method.

generation, through the values of the *frequency* and *sampling info* controls. *Append Wafeform.vi* was used to add the 0 value at the end of the generated signal in order to put the MOSFET in the blocked state so that the PV panel could be placed in the open circuit state at the end of the application execution. The fourth zone, (4), processes the measured signals and extracts the important parameters of the PV panel, such as: $\boldsymbol{P_{max}}$, $\boldsymbol{I_{max}}$, $\boldsymbol{V_{max}}$, $\boldsymbol{I_{SC}}$, and $\boldsymbol{V_{OC}}$. The ***R (V to I)*** constant is used for the conversion of the measured voltage across $\boldsymbol{R_{sense}}$ to the current. By multiplying the measured PV panel voltage and the determined current, the generated power is calculated. The I-V and P-V characteristics are determined and displayed in two graphs.

The same software application and load can be used to measure the I-V characteristics of the TEG in the case that a TEG replaces the PV panel and a temperature difference is ensured between its sides.

In [9], the implementation of the capacitor method is discussed. For better results, a more complex circuit was developed using an electromagnet relay for connecting the capacitor with the PV panel and a self-defined conditioning system based on AD8221 instrumentation amplifiers. For testing, a NI ELVIS II platform was used; see **Figure 24**. The LabVIEW application allows the comparison between data obtained through real measurements and the Multisim simulation directly in LabVIEW. On the measurement section, the actual application replaces zones (1) and (3) with one that controls the DIO channel, ***DO0***, which allows the activation of the electromechanical relay (**Figure 25**). The second zone remains almost the same with the difference of introducing the *DAQmx Start Trigger.vi,* which sets the start of measurements on an analog input channel only when the DO channel sends the relay activation command. Being a polymorphic VI, it was configured on the *Start Analog Edge* option from its dropdown menu. The VI inputs are set on the first channel as a source, rising slope, and 0.05 as level. When the relay is activated, the capacitor being fully discharged, it starts charging, and therefore, the current through the circuit passes from zero to short circuit value, which triggers the measurements. Using a *Case structure*, the application can be used with or without trigger. The application has an event-based architecture implemented with the help of an *Event structure* and a *While loop*. The defined event cases are:

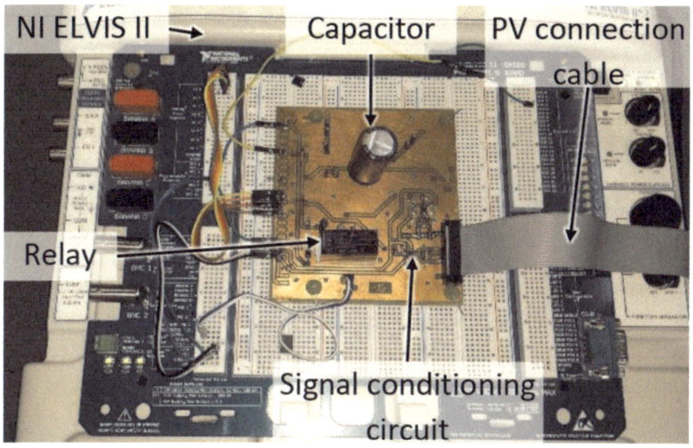

Figure 24.
The circuit for the capacitor method.

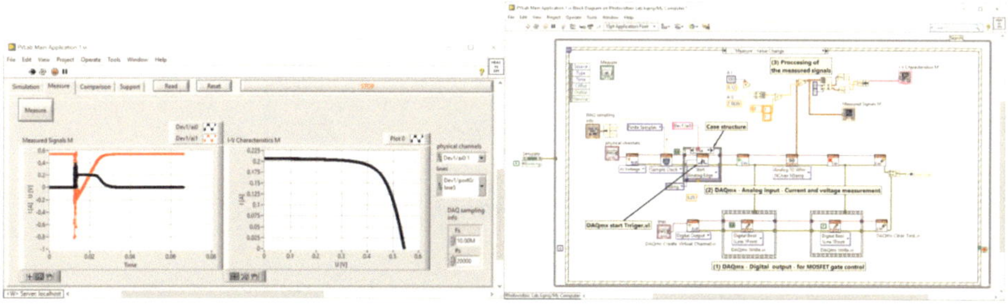

Figure 25.
The LabVIEW application for the capacitor method implementation.

1. ***Measure*** – allows measuring the I-V characteristics of a real PV panel;

2. ***Simulate*** – allows obtaining the I-V characteristics of a simulated PV panel;

3. ***Analyze*** – dedicated to comparing the real and simulated data;

4. ***Read*** – allows importing data from a previous saved file;

5. ***Reset*** – used for clearing all data for a new experiment;

6. ***Stop*** – clears all tasks and stops the application.

John Paliotta, "Software Quality and the Industrial Internet of Things: Why It Matters NOW," Embedded System Engineering, July 6, 2015. The RELab board dedicated for characterization of three RES is described in the paper "Design and implementation of RELab system to study the solar and wind energy" published in the *Measurement* journal [10]. The developed board is a modular one and can be used with three different devices as follows: NI ELVIS II, NI myDAQ, and Ni myRIO (**Figure 26**). By changing the board modules, the purpose of the RELab board can be configured for studying PV cells, small wind turbines, or small solar thermal panels. The software application was developed as an open LabVIEW project including some VIs as laboratory applications grouped on the three RES. The platform allows performing more than 30 lab experiments, out of which 21 are for PV cells, 6 for wind turbines, and 5 for solar collectors. An example of a lab experiment application is shown in **Figure 27**. The main architecture of the lab application is an event-based one. The user interaction with the user interface decides which state is executed. For example, by pressing the buttons ***Start***, ***Open***, or ***Save***, the application will execute the corresponding state, but also a specific state will be executed if the cursor of the ***I-V Characteristic*** graph is used. At the beginning, the application has a section that starts a web browser in the user interface, loading the lab work web page. In this manner, the user has access to the theoretical aspects of the lab and also the steps to follow to perform the laboratory work. In the measurement state called *Start*, there are three lines dedicated to reading the data from the PV cell and sensors. The first line reads the current and voltage generated by the PV cell and also its temperature based

Figure 26.
The RELab platform.

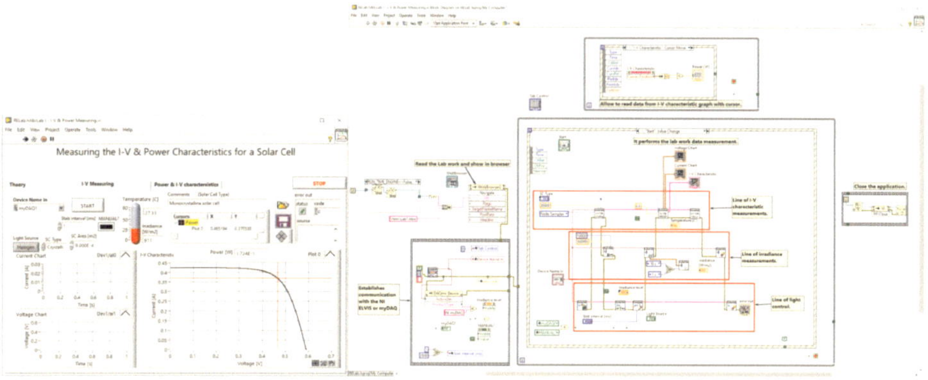

Figure 27.
A lab experiment application for RELab board.

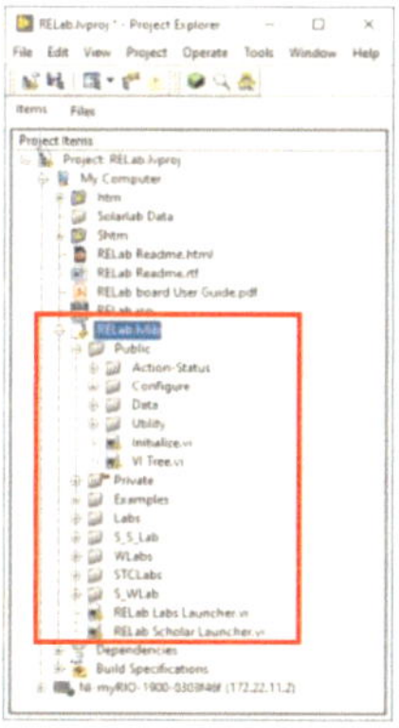

Figure 28.
A lab experiment application for RELab board.

on analog inputs. The second line has the purpose of reading the data from the irradiance sensor. This line uses an analog input of the NI ELVIS platform and a counter input for NI myDAQ and NI myRIO. The third line has the goal of controlling the level of irradiance, done based on varying the voltage level generated on an analog channel.

For an easier programming, the LabVIEW project was developed on a driver structure as can be seen in **Figure 28**, highlighted with a red frame. Based on the included subVIs, the user can develop new lab work applications based on their own concepts and methods.

3.2 LabVIEW and NI cRIO

For complex applications, a characterization system was developed along the NI cRIO 9074 platform. The system has four independent channels so that it allows the characterization of four synchronous RES. A self-developed electronic load with four independent channels was developed based on the capacitor method. NI cRIO is an industrial hardware dedicated for measurements, monitoring, and industrial control.

The NI cRIO 9074 has the following characteristics: 400 MHz CPU, 128 MB DRAM, 256 MB Storage, and 2 M Gate FPGA. The platform can act as an embedded system running applications for control and monitor without a PC connection.

The LabVIEW project is shown in **Figure 29a**. It can be seen that there are three levels of the project:

1. My computer branch (PC level) – which contains applications that are running on the PC. The *PC Main v1.vi* has the goal of showing and sending the necessary parameters to the embedded system. Also, the application accordingly processes the data received for I-V characteristics, extracting the main parameters as V_{OC}, I_{SC}, P_{max}, V_{max}, and I_{max} for all four channels.

2. NI-cRIO9074 branch – Real Time (RT) level – dedicated to running the application on the RT processor of the NI cRIO platform. The *RT Main V1.vi* takes data and sends data to the FPGA application and also stores the data on the internal permanent memory under TDMS format.

3. FPGA Target branch – FPGA level – contains the applications that are running at the FPGA level. The *FPGA Main v1.vi* has the goal of data acquisition from the analog input channels (current, voltage, and temperatures) and generates digital states for enabling/disabling the electronic load channels.

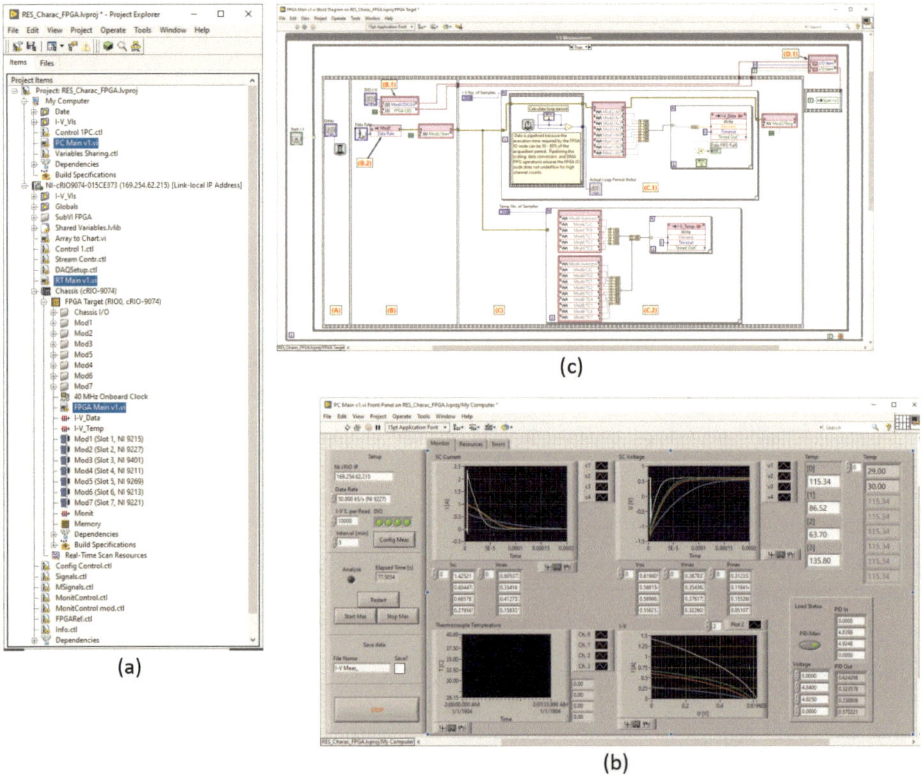

Figure 29.
NI cRIO application, a) LabVIEW project; b) PC user interface; and c) the block diagram of the FPGA application.

DOI: http://dx.doi.org/10.5772/intechopen.110298

The block diagram of the FPGA application is shown in **Figure 29c**. The application runs continuously using an infinite while loop containing a case structure that allows controlling the start of the measurement – which happens if the **Start I-V** variable receives the True value from RT application.

The *Sequence* structure allows controlling the steps of measurements. In the first frame (A), a settable delay is introduced to make sure that the RT time application passes to the monitoring state for receiving data. Frame B sets the used channels and also activates the user LED of the cRIO platform for knowing the state of the application, with the debugging purposes (B.1). Also, in this frame, the acquisition rate and the starting of the acquisition are settled (B.2). In Frame C, the data acquisition is done for current and voltage using the loop C.1 and for temperature, the loop C.2. The acquired measurements are transferred to the RT application using two FIFO queues: **I-V_Data** and **I-V_Temp** (see **Figure 29a**). The transfer is based on the DMA technique. When the required number of data is achieved, the data acquisition is stopped, and the used channels and the user LED are disabled (D.1).

The code of the RT level application is shown in **Figure 30**. The code is based on two while loops: first ensures the communication between this application and the application running on the PC (**PC-RT Loop**), and the second is dedicated to controlling the entire system (**RT-FPGA Loop**). This loop implements a state-machine architecture, which has the following states: **Idle**, **Init**, **Config**, **Monit**, **Status**, **I-V**, **Data Save**, **Stop**, and **Restart**. The **Init** state deploys the bitfile (A) obtained by compiling the FPGA application. At the same time, a TDMS file with a predefined name is opened (B). By reading the configuration shared variables in the **PC-RT Loop**, the running parameters of the RT and FPGA application are fixed in the **Config** state. The I-V characteristic is measured at a predefined interval of time in the **I-V** state (**Figure 31**). The line (A) is dedicated to the starting DMA transfer using the

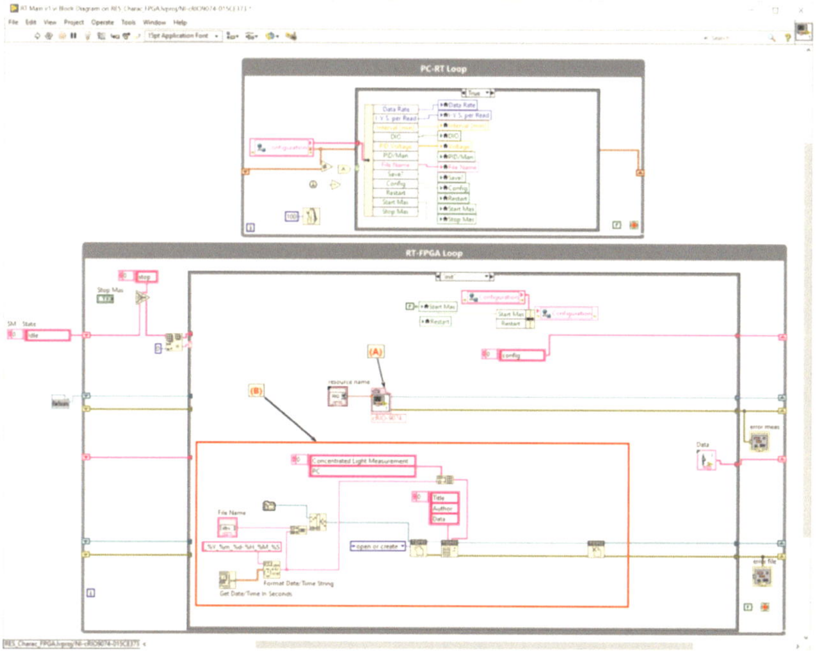

Figure 30.
The block diagram of the RT application.

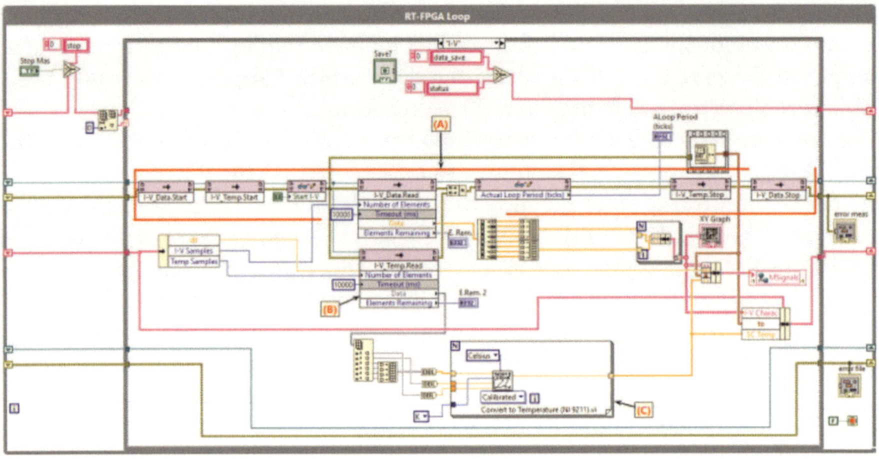

Figure 31.
The I-V state of the RT application.

FIFO queues. By sending the true value to the **Start I-V** button on the FPGA application, the data acquisition is started; then, through the *I-V_Data Read* (A – line) and *I-V_Temp Read* (B – line) nodes, the measured data are transferred from the FPGA to the RT application. Within the (C) loop, the read voltages from the K-type thermocouples are converted to temperature, in the Celsius scale. The read data are transferred to the PC application through the shared variables Measured Signals (**MSignals**) and through **Data** cluster to the **Data Save** state for storing the data in the opened TDMS file.

The shared variables are programmatically called in the PC application (see **Figure 32**) using the appropriate path and the Shared Variables subVIs (*Functions> > Data Communication > Shared Variable*): *Open Variable Connection*, *Read Variable*, *Write Variable*, and *Close Variable Connection* (**Figure 33**).

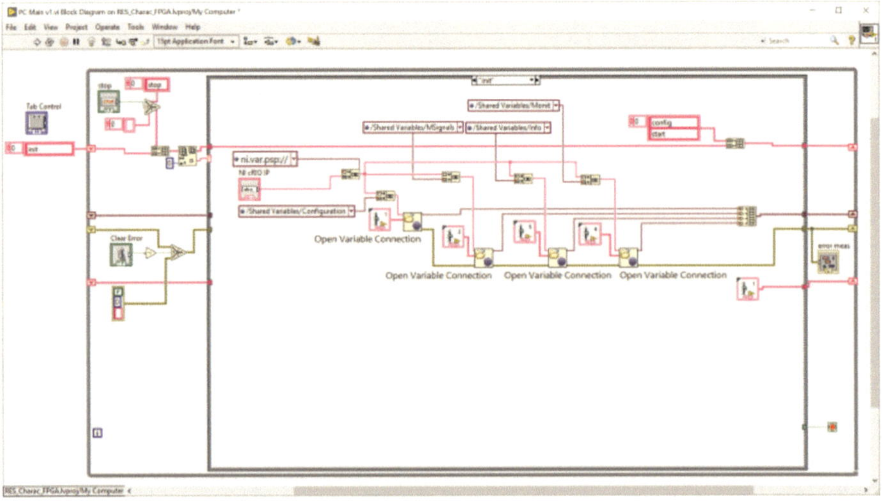

Figure 32.
*The **Init** state of the PC application.*

Figure 33.
The shared variable palette.

After the application is started, even if the PC is disconnected, the NI cRIO platform continues to measure and store data based on the defined parameters. When the PC is reconnected, the application continues to run without any interruption.

The saved files on the NI cRIO platform can be downloaded using the FTP protocol using File Explorer or any other FTP client.

4. Monitoring and industrial application

This section describes a real-life application, where a dedicated monitoring system is deployed on a PV-TEG to understand its benefits when supplying a household, from both the energy savings (avoiding producing energy from conventional sources) and the power quality perspectives. Once data is collected and processed, the next step is to publish data by means of the IIoT.

IIoT develops the central nervous system of a smarter world, building on equipment and devices that can measure and interact with humans [11]. According to the IEEE Standards Association, IIoT will become one of the drivers of growth in a wide range of technologies, with huge business potential, valued at $ 14.2 trillion in 2030. Within IIoT, data communication follows three distinct paths [12]: machine-machine (M2M), man-machine (H2M), and machine-smartphone (M2S), where the smartphone can be any other device with a touch interface, for example, a tablet.

4.1 System's structure

The deployed PV-TEG system contains 16 monocrystalline JA Solar 340 W panels, connected to a Fronius Primo 5.0 power inverter, generating up to 5.5 kW, as seen in **Figure 34**. This is the so-called *prosumer* configuration, where no storage unit is installed, so the energy excess is delivered to the power utility provider.

Figure 34.
Deployed PV panels on the household's roof, Google maps and photo.

The implemented measurement application has the main objective of gathering and exposing the process data according to the IIoT concept, which ensures visibility to various users. Using cloud technologies, the system achieves higher performance rates in terms of reliability and security. An alternative which has been considered would have used an on-premises solution, but the latter one raises additional costs and has a higher degree of sophistication, with a lower efficiency in operation.

Regarding the measurement hardware, the most important criterion considered in this project is its performance, that is, processing power, memory, high-quality DAQ modules, programming capabilities, communication interfaces, and internet integration.

The chosen solution is based on the NI Compact RIO platform, being deployed on two systems using NI cRIO 9033 controllers, which support NI Linux RT. The cRIOs 9033 are equipped with voltage (400 V RMS) and current (50 A RMS) modules, among other dedicated ones, as seen in **Figure 35**.

The investigated system has several operating requirements, influencing its design: (i) 24/7 operation; (ii) autonomous operation; (iii) easy to program and maintain; (iv) easy to expand without significant impact on the measurement system's performance; and (v) data access: web based, secured.

The application is multilayer type, distributed, according to the methods represented in **Figures 36** and **37**, respectively. NI LabVIEW™ is the software development environment used throughout the entire software implementation:

- Primary level of data acquisition, DAQ. The implementation is at the FPGA level, facilitating the collection of data at higher speeds, so that the data analysis support is provided at the upper, RT level

- Level of analysis of process data: (i) aggregation of data collected at the FPGA level, high-performance communication routines with the FPGA infrastructure; (ii) synchronization of collected data; (iii) effective data analysis: total harmonic distortion (THD), phase shift; and (iv) calculation of process quantities: effective values, frequency, powers, phasors, etc.

Figure 35.
NI compact RIO in use, monitoring the PV-TEG.

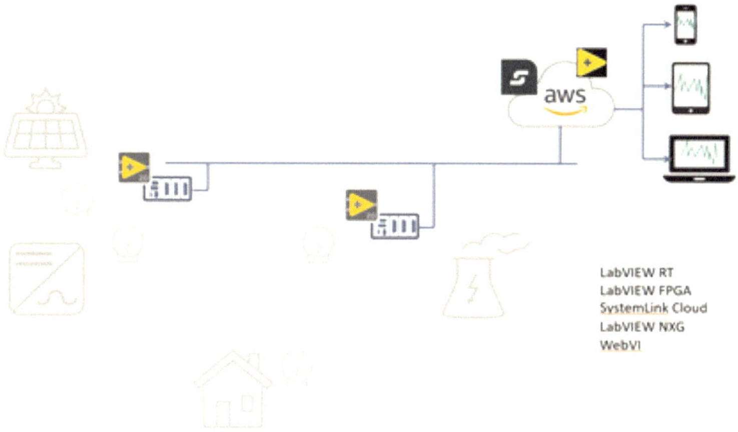

Figure 36.
PV-TEG with its associated monitoring system.

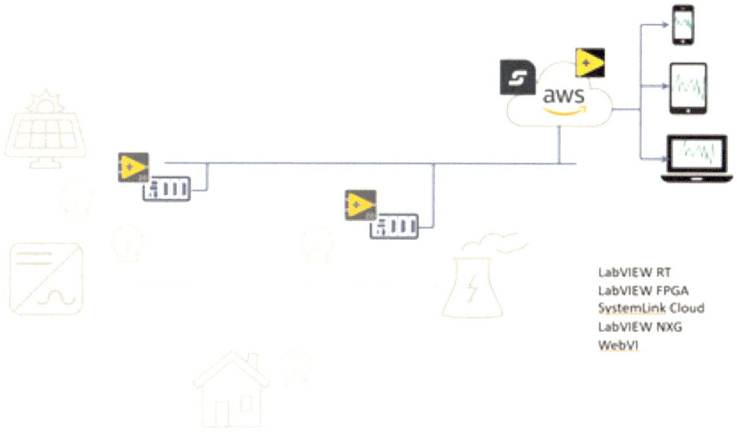

Fig. 36 PV-TEG with its associated monitoring system

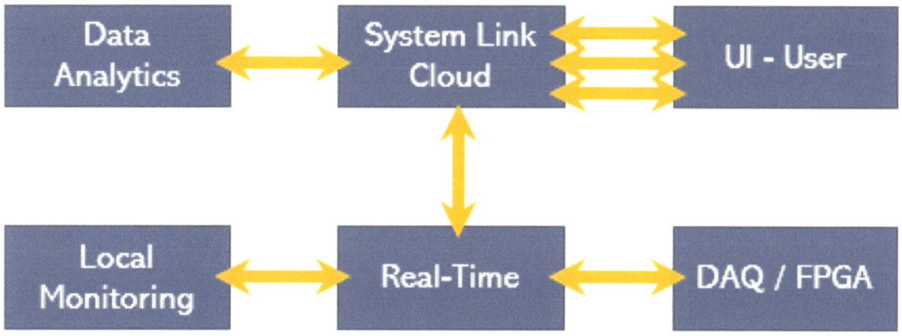

Figure 37.
Multilayer SW architecture of the proposed system.

- Local display of data for monitoring and maintenance
- Primary redundancy level implementation
- Additional analysis support
- Secured communication with the cloud application
- Communication with field equipment using the OPC UA protocol, mainly system services

At the top level, the application contains several modules:

- Data server module: definition, configuration, and operation of tags (process data)
- Cyber security module: defining the keys and security policies that address the communication with the primary measurement equipment
- User interface module: User-exposed application development using high-performance toolsets
- Data analysis module: stand-alone, workstation-level implementation: cloud interface, advanced analysis, post-data processing, statistical and process optimization, and cloud data exposure to facilitate third-party access.

4.2 Data acquisition software

The reconfigurable FGPA section is in charge of the high-speed data readings from sensors, passing these data along to the next level, the Real Time part of the application, by using dedicated FIFO mechanisms, as seen in **Figure 38**.

The Real Time part, the connected instance to the primary DAQ-FPGA loop, manages the lower speed part of the application, dealing with: (i) data retrieval from the DAQ FIFO, served by the FPGA app; (ii) data analysis related to the electrical specific values: power calculation – active, reactive, power factor, total harmonic distortion, efficiency, etc. Extra analysis functions could be easily added, with almost no penalty to the computing performances. Data retrieving section could be seen in **Figure 39**. Data collected from the FPGA primary source is copied to a shared variable structure, which handles the access of the other sections of the software.

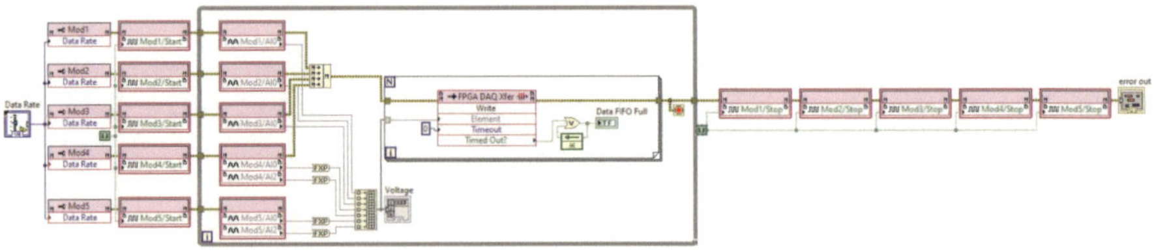

Figure 38.
DAQ implementation, FPGA level.

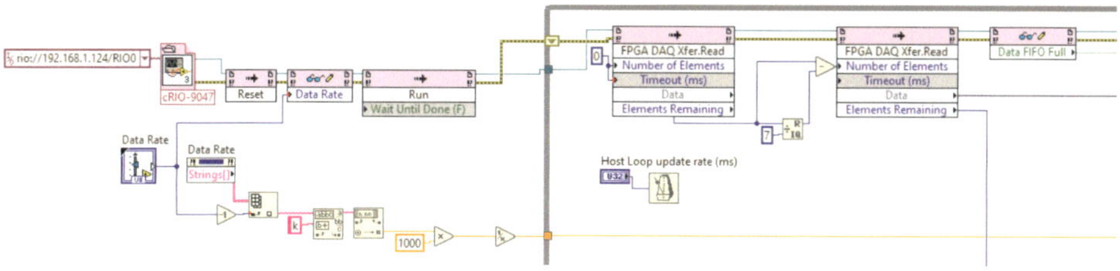

Figure 39.
DAQ implementation, RT level.

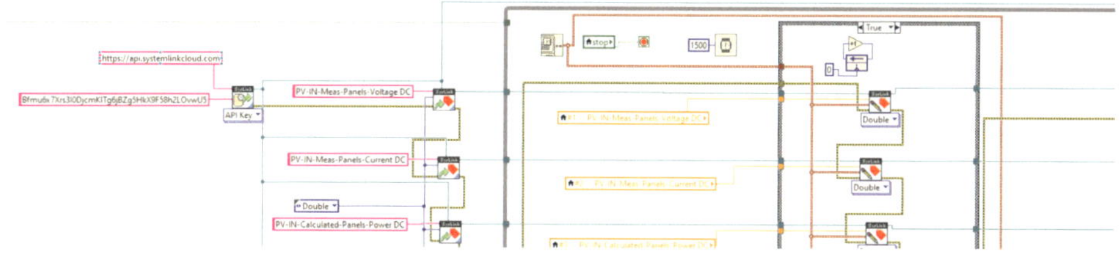

Figure 40.
Cloud communication protocol implementation, RT level.

Just to add more details to the previous statement, **Figure 40** depicts the communication implementation toward the AWS cloud using the NI SystemLink Cloud solution.

In this figure, one can see how shared variables, which support the communication protocol between different pieces of the software implementation, are pushed up to the NI SystemLink Cloud implemented using AWS support.

4.3 Cloud-based user interfaces

Figure 41 captures the user interface, which serves the embedded measurement system. One can observe the project window (left side of the left screen), accompanied by the main dashboard screens, which contain process data plots – raw data from system sensors, calculated values (power, power factor, efficiency, and financial data), plus other parameters describing the system's health. All this information is live data, allowing the user to efficiently manage the system, supporting business decisions, and also serving as a maintenance tool.

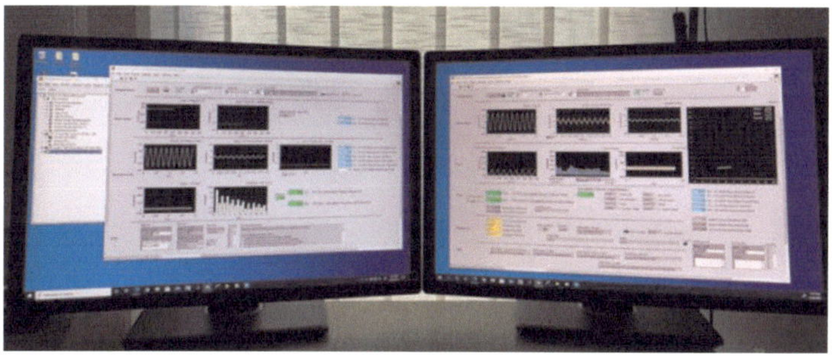

Figure 41.
DAQ user interfaces, system monitoring, data analysis and maintenance.

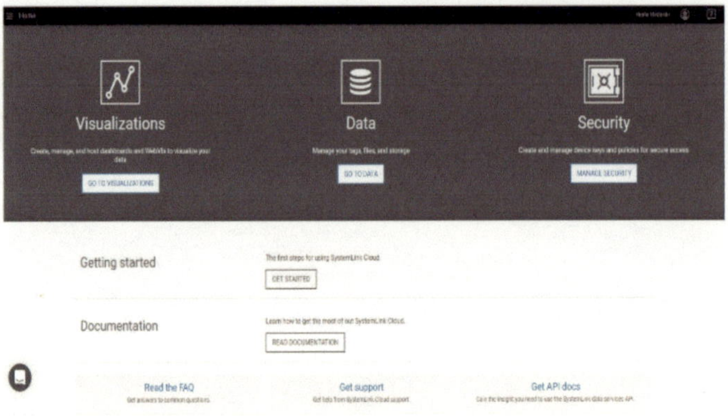

Figure 42.
NI SystemLink cloud landing page.

When designing the monitoring application, the final part of the software development has to manage the process data received from the field equipment. PV-TEG process data pushed to the cloud is organized in so called 'tags.'

The NI SystemLink Cloud server benefits of several built-in capabilities, see **Figure 42**: Data *Management* (tags – process information), *Cybersecurity* – API keys and policies, *User Interfaces* – various templates with more or less displaying capabilities, or user-defined interfaces developed using the state-of-the-art WebVIs created using LabVIEW NXG.

The LabVIEW NXG UIs are a more refined way to display process data coming from field. An important advantage is the capability of processing information through the LabVIEW diagram, whereas the other options are display only. As seen in **Figure 43**, data retrieved from NI SystemLink Cloud is processed (power-related math, financial info, THD, etc.) and displayed, either as instantaneous values or as waveforms.

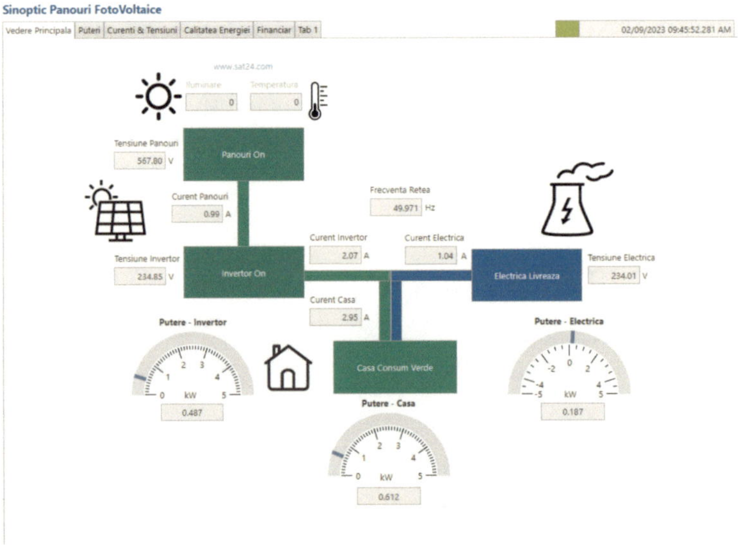

Figure 43.
Cloud-based UI, LabVIEW NXG diagram.

Such examples are shown in **Figures 44-46**, where one can contemplate a general view of the PV-TEG-based power generation system, containing instantaneous information describing the system's behavior (voltages, currents, powers, etc.), followed

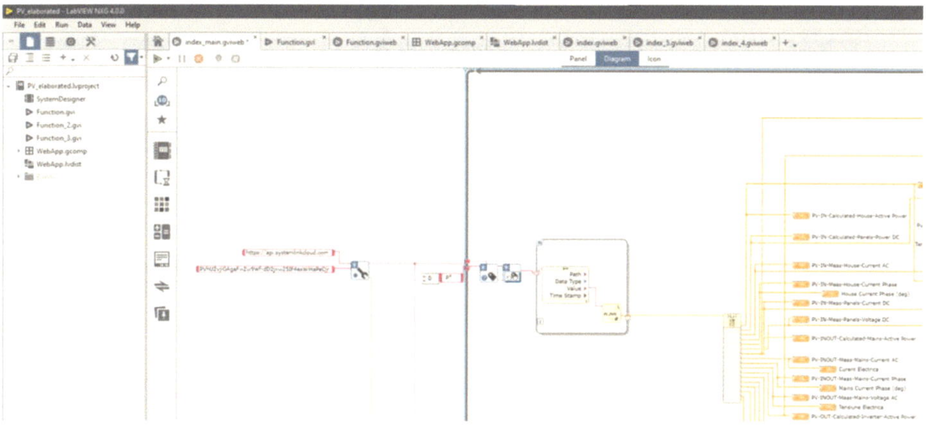

Figure 44.
Cloud based UI, LabVIEW NXG diagram.

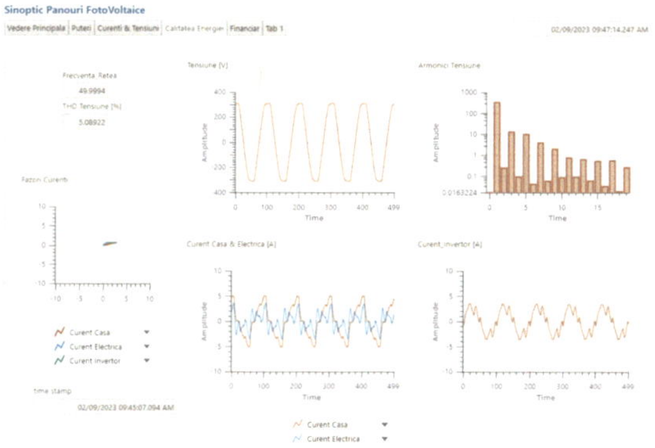

Figure 45.
Cloud based-UI, electrical power quality tab view.

Figure 46.
Cloud-based UI, financial daily report tab view.

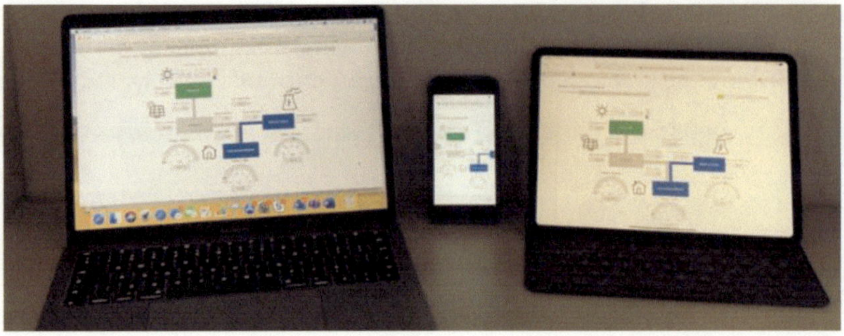

Figure 47.
IIoT NI SystemLink cloud-based interfaces, sys monitoring, data analysis, running on portable and mobile devices.

by a series of other screens where different trends are displayed, depicting the dynamics of the monitored system. All these parameters are retrieved from the cloud DB NI SystemLink Cloud, making the whole development process streamlined and effective. As illustrated in **Figures 45** and **46**, real-time data of selected parameters serve as a powerful analysis tool that allows the user to observe waveforms and to perform electrical power quality and financial analysis on raw and processed data.

The NI SystemLink Cloud [13] implementation brings in consistent advantages for data security and access. User Interfaces are tailored in such a manner that fit different devices, starting with computers and ending with portables (smart phones and tables), as shown in **Figure 47**.

The PV-TEG monitoring application is live, the access is granted to all interested users by simply accessing this link: https://hosting.systemlinkcloud.io/webapps/72a70649-5148-485e-b689-04b859fd1cf5/content/ApplicationFiles_64/index.html

Compared with a classical web server implementation, the cloud solution serves as *de facto* IIoT approach, being scalable, easy to maintain, and very efficient in information dissemination to a large spectrum of users, regardless of their technical abilities.

Besides the technical/engineering data, financial information is also available, as per user needs. These numbers play a crucial role, of course, in assessing the overall effectiveness of the PV-TEG system while running.

5. Conclusions

This chapter presents different solutions for implementing different simulations for PV and TEG modules, real characterization of the PV and TEG modules based on real hardware, and a solution for implementing a real monitor system for a photovoltaic home application.

Starting from the fact that the energy demand has increased worldwide and therefore new energy sources or increasing the efficiency of the existing ones is necessary, the RES has become very important nowadays as clean solutions instead of fossil energy sources. Therefore, the study, the understanding, and the usage of the RES are very important, and also it is very important to have solutions for training the personal for implementation and maintenance of the RES. Using LabVIEW along with other tools and hardware devices represents a very versatile solution.

LabVIEW allows different approaches for RES simulation through direct coding, using simulation toolkit or in combination with other software applications like NI Multisim.

Interacting with real RES is very facile in LabVIEW by using the very powerful API for data acquisition and the appropriate data acquisition hardware, from simple data acquisition board, like NI USB 6215 to NI ELVIS platform or more complex NI cRIO.

The real-life application presented in this chapter stands for a rapid prototyping measurement system using embedded equipment instrumented with graphical programming tools, starting from lower-level operation (DAQ), all the way up to its User Interfaces deployed in Cloud. The integrated way of data manipulation, from sensor to monitors and computer displays, allows the users to learn efficiently about the system's behavior and how to improve its performances while using.

All these collected data could serve for future analysis and optimization using more sophisticated techniques, like artificial intelligence, which could easily connect to IIoT-based solutions developed with LabVIEW™.

Author details

Petru Adrian Cotfas[1*], Daniel Tudor Cotfas[1] and Horia Hedesiu[2]

1 Electronics and Computers Department, Faculty of Electrical Engineering and Computer Science, Transilvania University of Brasov, Romania

2 Electrical Machines and Drives Department, Faculty of Electrical Engineering, Technical University of Cluj-Napoca, Romania

*Address all correspondence to: pcotfas@unitbv.to

References

[1] Herrmann M. The Global Population Will Soon Reach 8 Billion—Then What?. Available from: https://www.un.org/en/un-chronicle/global-population-will-soon-reach-8-billion-then-what

[2] Wole-osho I, Okonkwo EC, Abbasoglu S, Kavaz D. Nanofluids in solar thermal collectors: Review and limitations. International Journal of Thermophysics. 2020;**41**:157. DOI: 10.1007/s10765-020-02737-1

[3] Cotfas PA, Cotfas DT. Comprehensive review of methods and instruments for photovoltaic–thermoelectric generator hybrid system characterization. Energies. 2020;**13**(12): 6045. DOI: 10.3390/en13226045

[4] https://www.ni.com/ro-ro/shop/electronic-test-instrumentation/application-software-for-electronic-test-and-instrumentation-category/what-is-multisim.html [Accessed: July 2022].

[5] Cotfas PA, Cotfas DT. New Models for Photovoltaic Cells in Multisim, published by W. Mahmoud, Available from: https://forums.ni.com/t5/NI-Circuit-Design-Community-Blog/New-Models-for-Photovoltaic-Cells-in-Multisim/ba-p/3473652?profile.language=en [Accessed: July 2022]

[6] Cotfas PA, Cotfas DT, Machidon OM. Modelling and PSPICE simulation of a photovoltaic/thermoelectric system. IEEE 22nd International Symposium for Design and Technology in Electronic Packaging (SIITME). Oradea, Romania. 2016. DOI: 10.1109/SIITME.2016.7777272

[7] Bmac, Solar PV Power Converter and I-V Characterization Examples. 2014. Available from: https://forums.ni.com/t5/Power-Electronics-Development/Solar-PV-Power-Converter-and-I-V-Characterization-Examples/m-p/3442913 [Accessed: August 2022].

[8] Kaplanis S, Kaplani E. Renewable Energy Systems: Theory, Innovations and Intelligent Applications, D.T. Cotfas and P.a. Cotfas, Chapter IX: PV Innovative Techniques and Experimental Test Sets, Nova Science Publishers, USA; 2013

[9] Cotfas PA, Cotfas DT. Graphical system design approach in photovoltaic energy laboratories. In: IEEE 21st International Symposium for Design and Technology in Electronic Packaging (SIITME). Romania: Brasov; 2015

[10] Cotfas PA, Cotfas DT. Design and implementation of RELab system to study the solar and wind energy. Measurement. 2016;**93**:94-101. DOI: 10.1016/j.measurement.2016.06.060

[11] Paliotta J. Software Quality and Industry 4.0: The Industrial Internet of Things. DZone, 2015. Available from: https://dzone.com/articles/software-quality-and-industry-40-the-industrial-in. [Accessed: July 2022].

[12] Greengard S. The Internet of Things. Cambridge, Massachusetts, London, England: The MIT Press; 2015. p. 232. ISBN: 9780262527736

[13] Available from: https://www.ni.com/ro-ro/support/documentation/supplemental/18/systemlink–architecture.html